SECRETS OF THE FOUNTAIN OF YOUTH

SECRETS OF THE FOUNTAIN OF YOUTH

by
Mary Louise Zeller
and
Michael Akagi

Published May 2007 by:
KiwE Publishing, Ltd.
Spokane, Washington
http://www.kiwepublishing.com

Library of Congress Control Number: 2007932526

ISBN 987-1-933973-04-3

Printed in the United States of America

This book is dedicated
to everyone with an open mind.

Foreword

I am a tenth-degree Grand Master in Taekwondo and have been teaching this Korean martial art to U.S. Congressional Representatives and Senators on Capitol Hill for forty-two years. There are hundreds of Taekwondo schools around the world that bear my name and teach my curriculum. I am considered the father of Taekwondo in the United States and in Russia.

In August of 1993, along with another group of weary American world travelers, I arrived in the Inchon International Airport in Seoul, Korea. We were all guests of Dr. Dohyun Choe; a world famous healer in the Korean healing art of Su Ki. Dr. Choe is a Korean national transplanted for many years to Dallas, Texas, U.S.A.. Dr. Choe was the common mutual friend of all of us and had invited us to Korea for a rare event that challenged me to come… Superman Endurance Training. It was here I first met Mary Louise Zeller. I was delighted to discover that we had a mutual interest; she, too, was a black belt in Taekwondo.

During our visit, we had the opportunity to visit Taejon, Korea and the World Expo. As travelers often do, we began speaking to each other as a couple of Americans in another land, even though Korea is my country of birth. Mary Louise told me that she had started studying Taekwondo rather late in life, actually when she was forty-six years old. At the time we met, she told me she was fifty. She had just earned her second-degree black belt. As we talked, she told me that she was the current California State Champion in forms and sparring. I couldn't believe my ears. Here was a fifty-year-old woman, who looked like she was in her thirties, telling me she had just won two gold medals in a large state competition against women less than half her age! I know something about Taekwondo competitions, having put on major tournaments all over the world in this sport, and there was no other woman in the world accomplishing such feats at her age.

I invited Mary Louise to work out each morning while we were in Korea. As I ran with Mary Louise the first day, it quickly became apparent that she was in incredible shape. She had the energy of a teenager and appeared to be tireless. In the days that followed, my experience of her inexhaustible energy was repeated over and over, and I was convinced that for her age, Mary Louise was the most fit female martial artist I had ever met.

I subsequently invited Mary Louise to my home in Washington, D.C. and arranged for her to train with me and other Taekwondo masters in the area. She was an apt student and learned rapidly. During this time she told me of her aspiration to win a gold medal at the Olympics. I said to myself that this would be impossible for a woman of her age. The top women in our sport were much younger, and Mary Louise was getting older each year. Her "Impossible Dream" showed up for me as just that, an impossible dream. After a few weeks of coaching her, she left and began traveling the country with her husband and young son, and I didn't see her for several years. When I saw her again, I was pleased to see she actually appeared younger and stronger than before.

I was so pleased to see Mary Louise looking so well because I, too, am committed to being a model and example of strength, flexibility and vitality no matter how old I am. I stand for having the wisdom of a 100-year-old person in a body that looks like a twenty-one-year-old. You can actually join me in that commitment by joining my International 100/21 Club. I do 1000 pushups each day, and now in my mid-seventies I am as strong and fit as I ever was. Even after having a stroke at the age of seventy-two, following an open-heart surgery to repair a genetically defective heart valve, I have rehabilitated myself in just three years. I am just as quick now, in both body and mind, as before I had the stroke. Mary Louise and I are living proof that Taekwondo can actually reverse aging of the human body. It is my wish that the two of us will bring this message to the world and inspire thousands: "The body is not the limit. Aging may not be a choice, but the decline commonly associated with aging certainly is a choice. We will engage in the practice of Taekwondo throughout our lives to always have a body, mind, and spirit that can continue to do all that we love to do, all the days of our lives."

It is not easy to become an Olympian. It is even more difficult to win a gold medal at the Olympics. Mary Louise just keeps playing her game as if it were possible. Even if she doesn't win an Olympic gold medal in Taekwondo, at sixty-four years of age her journey is inspiring. She has won over 100 state, national, and international gold medals in Taekwondo competitions, all after the age of forty-six, with the preponderance of the medals won in her fifties, and now continuing into her sixties. Even as this foreword is being written, she is in Korea, training with the best in the world in Taekwondo technique, a new Olympic division. My bet would be on Mary Louise making the U.S. National Taekwondo Team in this new forms division. Mary Louise Zeller takes action consistent with her passion and commitments, not the number of years she has spent on the planet. She is truly ageless.

After you have read her story, you will never again look at aging in the same way. I invite you to allow yourself to be inspired as I have been by this remarkable woman and her story of achievement and go on to your own greatness.

I recommend that you create a vision for your life just as Mary Louise has. Contemplate what touches your passion and what is most important to you. There you will find your vision. Take action every day to realize your vision. Vision is life to the soul and to the body. If you want to be strong and healthy all of your life, challenge yourself to live as if you have a great purpose to accomplish. You do, if you create it. I have seen Mary Louise Zeller apply these principles in her life; now allow her to show you how to do the same.

Every day of your life, develop strength in your body, wisdom in your mind and honesty in your heart and your life will be a literal fountain of youth.

Grand Master Jhoon Rhee

Contents

LEFT-SIDE PAGES

Secrets of the Fountain of Youth

RIGHT-SIDE PAGES

Mary Louise Zeller's Story

Mary Louise Zeller is now a 5th Degree Master Instructor in the Olympic-style Martial Art of Taekwondo and has been involved in "youthing" (the art of growing younger) for some 16 years. The unusual aspect of her story is that she only began Taekwondo at the age of 46 as a 25-pound over-weight, mildly depressed, middle-aged housewife. She has been a 14-time U.S. National and 5-time International gold medalist in Olympic sport of Taekwondo. Now at the age of 64, Master Zeller still competes Nationally and Internationally and in 2004 she was inducted into the U.S. Martial Arts Hall of Fame, not as a coach like others of her age, but as "International Female Competitor of the Year". She has earned the nickname, "Ninja Grandma", and at her young age has recently trained at the U.S. Olympic Training Center in Colorado Springs, Colorado and the Kukkiwon International Training Center in Seoul, Korea.

Michael J. Akagi is a veteran marathon runner with over 30 marathons under his belt, including the prestigious Boston Marathon and is the fastest distance runner in his age group in Hawaii. He has been running for over 25 years and has made a lifelong study of the body as it relates to nutrition and conditioning. In addition to being an author, Mike Akagi, at 57 years old, is a high school cross country, track and running coach who leads by example and runs daily with his team. He has a black belt in Karate, is a popular motivational speaker, is a helpful husband, father, and grandfather of nine adorable grandchildren. "Youthing" is the theme of his life and his book, "Secrets of the Fountain of Youth" promises to deliver all the secrets that he has discovered over the years.

Preface

The purpose of this book is to reveal to you everything you need to know to become young again. These are the things that have escaped you, the things you have forgotten and the secrets you have never known. This book will give you the keys to unlock the doors that lead to eternal youth. Truly, you shall know all the secrets of the fountain of youth. *Secrets of the Fountain of Youth* will also allow you to see that "youth" is a state of mind. You will receive the knowledge necessary to view aging truly as a state of mind not as an absolute physical fact. By the time you finish this book you will know everything you need to know to unlock the doors that lead to physical well-being, emotional well-being, and looking and feeling younger than you ever thought possible.

Have you ever wondered why some people seem to age very quickly and why some people never seem to get old? Do you secretly want Peter Pan to be real? Do you know someone who is maybe 50 years old, yet looks and acts 25? You need to know that these people are not just a product of their extraordinary gene pools. Ageless vitality is not an accident of genes. It's a new mindset and practice. We are actually just thinking in a way that you don't and doing things you don't do. Is it time for you to start doing the things that can actually give you a youthful life all of your life? Do you dare to dream of great accomplishments and adventures at any age? Are you willing to do the work to have a body that can *do* those dreams at any age? You can have a life that you love, all the days of your life. You can then model that possibility and pass on that knowledge and mindset to others, your children, your grandchildren, and the next generation and the next. You can be a part of an evolutionary leap for human beings.

We have personally researched different phenomena, practices, and products that suggest that aging or "antiaging" is not solely a function of genetics. Most people feel that when someone looks young, it is because that person inherited a great set of genes. Hence, they live as if there is little or nothing that they can do to change that for themselves. They go on living their same hopeless, boring, old, and fat lives. Sounds too familiar? We offer you all the knowledge we have gathered through our combined 50 years of experience. You can decide for yourself whether you are willing to allow the common culture of aging to control you or if you are willing to invent a new possibility for yourself and others. The decline of aging is a choice! It may just be that the decline of aging is as much a state-of-mind phenomenon as are all the aspects of your existence.

We are going to show you that you can be in control of what is considered the "normal" effects of aging more than you ever thought possible. We will not only show you the living proof that will make you a believer, we will teach you the things you need to do to look, feel, and perform younger than you ever thought possible. We honestly believe that when you are finished practicing the secrets we have to reveal to you in this book, you will not only start becoming younger but you will actually be able to live a life that you love.

Maybe you remember the story of Camelot when young King Arthur used to hang out with Merlin, the magician. You may also remember that Merlin lived life backwards and therefore, could teach young Arthur much about how to live life. The secrets in this book are going to allow you to start living life much like Merlin did - backwards! Instead of aging, you are going to begin growing younger. We are going to call this process "youthing," not antiaging. Youthing is the art and process of growing younger.

Each day and each year, you will be getting younger because you will know the secrets that cause your body and your mind to grow younger instead of older. You will empower yourself with the ability to "youth" instead of age. Remember that aging is a process. Youthing is also a process that we, as human beings, spend very little time exploring or practicing. We have searched and researched to bring you the secrets of youth from all over the world. With the secrets we will reveal to you,

you will be able to unlock the door that leads to youthing. The opposite of aging is not antiaging - it is youthing. The reason why there is no word for it is that until now, nobody believed it was possible. What this book proves is that it is not only *possible* to grow younger, but you are going to do it! We invented this new word and the word is youthing and it means growing younger every day instead of older. Get used to the word. Put it in your vocabulary and start your youthing journey today.

You will be amazed at what happens to your body, mind, and soul as we reveal to you the secrets that allow you to unlock the keys of life that will enable you to keep becoming younger each year. Please keep in mind that the principles of youthing are a generational commitment and you can pass these things on to your children, grandchildren, and all those who are to follow. You need to understand that it may have taken your body several generations to get to its current condition. As sad as that may be, we can still make a difference for you and your body and your descendants no matter what state you may find yourself in today.

Each chapter of this book is written as a stand-alone entity. Therefore, you can read each chapter and determine for yourself if you need to use the secret of that particular chapter or pass it by or reuse it later. Likewise, in the future when you think that you have mastered the content of a particular chapter you can simply focus on another chapter. This will allow you to have control over every aspect of your new life.

We suggest that you read the whole book through the first time so that you get the full overview of all the things you need to know to begin and continue your journey to the new, younger you. Read the whole book one time through, then go back and focus on the individual chapters that you want to master, when you want to master them.

If you're ready, you can now begin one of life's most fantastic journeys. This journey is one that will transform your life in ways you could only heretofore dream about. Imagine having a life you love, full of the secrets of youth. You will be recreating yourself in a way that offers you the possibility of taking control of your whole life. ***Secrets of the Fountain of Youth*** is like no other book you have ever read. It is full of the intrigue of a mystery novel and so profound that you won't be able to put it down.

Not only will we give you all the secrets to a youthful life, we will give you information you didn't even know existed. If you have access to the Internet, we will give you access to a far greater world than you have ever known. If you don't have Internet access, we will inspire you to get it and learn it. All that we will reveal will have you on the Internet to find the products and information you can use to transform your life. Our web site, www.secretsofthefountainofyouth.com, will launch you into a new world where you can find and review all the secrets of youthing. You will want to go there often. One of the greatest secrets we will reveal to you is access to the information that allows you to be free. It is one of the keys, not only to youthing, but also to life. Time, and the wise use of time, is not only one of our secrets, it is what life itself is made of. Our first request is that you gain access to the Internet; it will save you a lot of time!

Now let the adventure begin.

Introduction

Being Unstoppable

The very first thing you need to do is to decide that no matter what happens you are going to stick with it. This is called being "unstoppable"! A simple explanation of this is that most people are easily stopped. Let me give you two examples of this. First, last Monday, I said I was going to go out and run four miles first thing in the morning. The first thing I did was get out of bed at 7:30 just in time to jump in the shower and run off to work which, coincidentally, was about four miles away from our house. Needless to say I never ran my four miles. My life was looking just like it usually does, busy and always full of things to do. No time to do the things I want or need to do to support my health and well-being. I was, stopped! We are frequently stopped from doing the things to which we are truly committed; the things that would make the most difference in our lives are the things that usually don't get done.

Next, my wife was supposed to take our youngest son to a birthday party several miles from our house. Much to my surprise, 20 minutes later she returned with our son and told me that I would have to take him because she couldn't find the right house. I discovered that she had been only two doors away from the birthday party. She was, stopped!

Just consider your daily routines. What do they look like? For most of us, our lives can be described by the phrases "same old, same old," "boring," "not again," "didn't I just do this?" "why me?" "I can't do this," and so forth. Isn't it about time something was put into place in your life that allows you to live life differently? Perhaps, what's missing or what you would really like to do or be in life can replace the previously mentioned "stoppable" words. Maybe some new words can start showing up, words that describe possibility like the words; *adventure, intrigue, excitement, fun, energetic, lit up*. You might be wondering how your tedious, boring life could undergo such a massive transformation.

What if we told you it was easier than you think, so easy, in fact, that you could begin this moment on a journey of exactly that; a journey of adventure, intrigue, excitement? You probably wouldn't believe us.

But here goes. Listen carefully. Life is whatever we say. It is not about having enough time or money. You have the power within you to create whatever it is you want life to be or whatever you want to do in life. If you can train yourself in being unstoppable, you can achieve anything you want and have life exactly the way you want it. What does unstoppable look like? How in the world could you go out and do whatever you decide you want to do, whenever you want to do it, no matter what your age? That is what this entire book is really about.

When you heard about this awesome new book or someone told you about this great new book that has some real answers to life, you wondered if it was really possible for anyone to write a book that promised to deliver the "fountain of youth." Not only is it possible, this is where the journey begins. Figuratively, you are going to have to strip naked, pack up the house, discard the old clothing, furniture, and baggage and literally throw it away. Then you are going to put on this set of incredible new clothes, pack a new bag full of the best stuff life has to offer, and head down a new trail full of adventure, excitement, energy, intrigue, and life the way you want it. Oh, and by the way, never look back, you might turn into a pillar of salt.

The key to our journey lies in being unstoppable.

What our little book and we are going to do is be your coach on this journey. We are going to give you the coaching you need to succeed. By the way, when one is being unpredictably successful, making a difference in the quality of life for all, and being true to one's ideals and dreams, one is being most alive! What that looks like is that anytime you need help through anything that is stopping you in life, you can turn to the particular part of this book that represents where you are being stopped and get the coaching you need to not just get you through it, but to break through and transform your life, again and again!

The most profound piece of coaching we can give you at this early point in your journey is that life is whatever you say. The very first step in this new journey is what you say about yourself and your life. In order to take the old baggage and throw it out the window, we are going to do some serious soul searching and identify some of your explanations about your existence and what you are saying about your life and growing older.

This book is designed like a series of treasure chests. Each chest has a key, so you will own a full set of keys when you master each chapter. Each key opens one of the precious treasured secrets of the fountain of youth. Each chapter contains a key. Each key is hidden and will jump out at you only when you understand it. We will not come out and tell you what the keys are, you must discover them for yourself. When you do find the key, it will open up a whole new world for you and literally become a treasure, a precious treasure from the *Secrets of the Fountain of Youth.* These keys are like secrets; they whisper to you and they become a part of you only when you spend the time and practice that generates mastery. Remember this is a journey; it begins with the first step and never ends. There is no arrival. Enjoy each step of the journey. It is fun. It is intriguing. It begins with reading this book and will end only when you stop engaging in the process.

The key element in being unstoppable is centered in your brain. You are usually stopped not by physical elements, but by what you think. The brain is so complex and so magnificent that we don't fully comprehend what goes on up there. Sometimes no, actually most of the time it controls us instead of us controlling it. The first step in youthing is getting in control of our thoughts or in other words, our brain.

Your brain is the most powerful and amazing piece of technology on planet earth. Do you understand this amazing tool? If your brain were a gadget that you could buy at Circuit City, you would be simply amazed at the technology. There is nothing quite like it in the whole world. The big question is, what are you doing with it?

The first part of your journey is being unstoppable in finding all the keys that unlock the secrets of the fountain of youth.

SECRETS
OF THE
FOUNTAIN OF YOUTH

MARY LOUISE ZELLER'S STORY

CHAPTER 1

THE BRAIN

Access to Power and Youthing

As we consider the process of youthing, what should become obvious is that the process of becoming younger, instead of older, starts with the mind and the brain. We have often told our children that one of the most important things to learn in life is that the brain is the center of who we are and that we need to learn how to control it and our thoughts. We have tried to explain the power of the brain to them. Likewise, we will try to reveal the power of the brain to each of you. A very simplistic way to distinguish the brain from the mind is that the brain is the biological "computer" of the physical body. It's designed to manage all the functions of life for the human body as well as gather and sort all information that pertains to the function of the human physical body. The human mind has a purpose also. Its purpose is survival for the human being. It is the human "spirit" accomplishing this by interpreting all of data gathered by the bio-computer. This data is held in a system of code, i.e., language or the fundamental element of language — words. The purpose of the mind is then to interpret data in order to maintain the survival of the being or of whom the being interprets itself to be. As you will see, it's that last statement that determines how our lives will go.

CHAPTER 1

NINJA GRANDMA

Worn Out to World Champion

Everyone loves a good story. I will be telling you my personal story, but this book is not about me — it's about you. The only thing you really need to know about me is how my story might inspire you and impact the quality of your life. That empowering you is satisfying for me, is really of no interest to you, except that as you expand you will realize that sharing with and empowering others will be satisfying for you too. Empowering people to live their lives in full self-expression and greatness is so rewarding it is almost seductive.

My upbringing was probably ordinary and middle-class. A few aspects, though, made lasting impacts on my life. We moved a lot and I went to 14 different elementary schools. A year in one place was a good year for me. We moved to Atlanta, Georgia when I was 14 years old and stayed there. My sisters were usually upset with the news that we were going to move again. They cried and were upset and sad for weeks before and after the move. They hated to leave their friends. I did too, but the sadness was

Many years ago I read a series of success motivation books. Perhaps you have read *Think and Grow Rich* by Napoleon Hill or another book by W. Clement Stone written in conjunction with what Napoleon Hill called *Success Through a Positive Mental Attitude.* The key statement of both of the books is: "Whatever the mind of man can conceive and believe it can achieve." Think of the power of that statement. Anything you can get your brain to believe, the brain can achieve. One of the processes or one of the keys that is absolutely necessary for your youthing journey is to convince your brain that the things you think are actually possible to achieve. Then give your brain all the things it needs to take you on this incredible journey. The brain is that powerful. It will deliver.

Here is an amazing explanation of the power of your own brain and some ideas on how to direct or harness this limitless power. Understanding your own power and the power you possess as an individual is one of the biggest secrets that we can reveal to you. You can start believing not only in your brain, but also in yourself, and realizing that you truly are as young as you can get yourself to believe. When you fully understand how important this truth is, you will begin to understand what this chapter is all about. Your brain is more powerful than all the computers in the world combined. You sit there and think, oh sure, that can't be true. In *Think and Grow Rich*, Napoleon Hill talks about the brain and how to get it to be within your control, instead of it controlling you. That is an interesting proposition, the fact that you could actually be in control of your brain and of your life.

What would happen if you are able to do that? When you are actually able to be in charge or in control of your thoughts, your whole world will transform; your whole being will transform. You can literally start on any journey you want, including youthing, but you will not be limited to just youthing. You can become or have anything in life you want and you can do that no matter what your age. You can literally transform your life, at any time of your life, and have it exactly the way you want. What an incredible thought — have life exactly the way you would like it.

First of all let's explain to you why your brain is so powerful. The brain is equivalent to a computer, right? It is a master center of data and information. How massive is it? Well, just to begin to understand, blunted by my excitement about travel and the adventure of a new place. The way my sisters dealt with the upset of continual change in their childhood was to stay in the same place their entire adult lives. I have continued the adventure of change and traveling to new places. It not only gave me a hunger to experience more in life, it seemed to have made me sturdy enough to travel the world, even by myself. I have always been willing, even sought to try new things and have extraordinary adventures. I sometimes say I don't have time to die-I still have too much to see and do. My family tells me I must have gotten the wanderlust gene from my father. They've gotten postcards from me from all over the world. I now live in one of the most extraordinary environments in the entire world and even though I still love to travel, I plan to live in the mountains of Sundance, Utah for the rest of my life. Although I recommend travel as an experience that greatly expands your view of life, and thus expands you, I've finally found my place in this world and I'm digging in and making it mine.

Another impact from my early years that stuck with me all my life is the love of sports and physical activity. Maybe this personal expression in sports was initially my way of getting my father's attention. It's interesting how our preferences and personal styles are formed. My older sister was captain of the cheerleaders, Homecoming Queen, and had the bubbly personality of Doris Day. It didn't seem to me that I could compete with that. Bonnie did that one too well. I spent my time out in the woods building forts and catching polliwogs. I was the third girl. As soon as my father saw me, one more girl, he never paid much attention to me. The only activity of mine my father came to was a sports event. My father loved how I scrapped for the ball when I played point guard for my high school basketball team. He was the loudest voice in the stands. I think it was the only positive attention I got from him.

Napoleon Hill says in his very first chapter that the brain and the communication lines that interconnect all the parts of the body are massive. If you take those lines and compare them to an actual computer, the data is astounding. The lines that inter-connect your brain and all your body would be equivalent to a three followed by one million zeroes — not the figure three million, but the figure of a three followed by one million zeroes. I believe that is called a google. So your brain and body are connected by three google lines. This is phenomenal. It is no wonder that the brain is so powerful. Your brain is more powerful than all the computers in the entire world linked together in one super computer. Please take a moment and think about the power that lies for the most part dormant, but is within your ability to access and utilize.

Let's talk in terms of what the brain can do. The brain can literally bring you anything you want in life. It can deliver to you anything you want. It is almost scary. Whatever you keep in front of your mind or whatever you think about the most, the brain delivers to you. If you will review your life for just a moment; everything that has transpired in your life (good or bad), anything you own, anything you have, anything you have mastered, any part of knowledge that you have in your brain, is a result of the brain delivering it to you because it is what you have kept in front of your mind. Literally, whatever has occupied your thoughts shows up. If you have a thought that you aren't good enough or that you don't deserve something, whatever it is, it has come about (negative or positive) as a result of your having those thoughts. We highly suggest that you begin to feed your brain with positive thoughts and declarations that will bring you everything you desire in life. The most important thing is to surround yourself with thoughts of wellness and youthing.

A True Story

Once, many years ago, I think it was in 1981-82, after having read *Think and Grow Rich* by Napoleon Hill and *Success Through a Positive Mental Attitude*, I set up a series of goals. Part of the advice and direction from those two great books was that you have to keep what you want to have or what you want to own in front of your brain. It's very interesting that the two times that your brain is the most susceptible to ideas and information is just before going to sleep, and when you are watching

He's not here now, and I still love to play games. I am competitive. One of the things I always say is that in order to feel fully alive, I have to be moving my body. I am not a happy camper unless I am sweating and engaged in some competition, adventure, or project that is challenging me to learn, practice, and grow. I'm 63 years old and that has not changed at all. Maybe the only change has been that the size and scope of the games and adventures have gotten bigger. By the time many Americans become adults they have substituted being excited about watching games rather than playing games. It is fun to watch games, but watching is entertainment, playing is enlivening. Watching ultimately makes no difference. Playing makes the critical difference. It may seem like a bold assertion, but I'll dare to assert that one of the elements of "the fountain of youth" is playing physical games and engaging in the game of mastery. I've read of a place near India called Aureville. It's located in a remote place in the Himalaya mountains. The people who reside there are reported to live well over 100 years. This article claimed that men even fathered children when they were over 100 years old. One of the aspects of their remote civilization was that they not only had a pure natural diet, but also played physical games every day. They loved the aspect of competition. It is reportedly a big part of their social structure.

Some people think competition is bad. I've heard them say it's bad because there is always a loser and that damages self-esteem. I propose that quite the reverse is true. Even though there is a loser and a winner in competitive games, both losing and winning have benefits to the human being. Losing shows us where we need to improve; it strengthens us to bear disappointment with the grace of maturity rather than the grief of a child. Losing can strengthen our resolve to focus in our endeavor and master it. In losing we have

TV. What we suggest you do is quit watching TV and second, start feeding your brain just before you go to sleep and the first thing when you wake up in the morning. Whatever it is you want to achieve in this life is what you need to express in words and keep in front of your brain. Goals are a good example. You need to keep the very things you want about youthing in front of you. You need to keep the concepts and goals you have associated with your vitality in front of your powerful computer and let it take you where you want to go.

Let me finish my experience from 1982. I decided to write up my goals in such a way that my brain would understand exactly what I wanted. This way you can understand what a miracle this whole thing was, I need to describe my family and the financial circumstances under which the formation of these goals began. First of all, we were financially strapped. I owed approximately $80,000 to $100,000 to various creditors. Life was a mess. It was a struggle. I left my full time job because I wanted to be an entrepreneur. I really wanted to start my own business. I started a landscaping business. During this time of great pressure and wondering what to do in life, we were blessed with two little children. For those of you who have children or had children, you know it is a very difficult time of life when your children are small. As I was in the middle of these great tribulations, I decided to start reading the two motivational books I mentioned, *Think and Grow Rich* and *Success Through a Positive Mental Attitude.*

I was intrigued by the power of the brain and what the books claimed it could bring to me in life. I decided to test those principles out. I set a series of goals, very specific in nature, and decided to follow the directions of the book and read those goals to myself every morning and every night. It really seemed dumb at first, and I thought there wasn't any way any of those goals would or could really happen. In the beginning I really didn't believe, but remembered that the books said, "Whatever the mind of man can conceive and *believe* (that is the key the word, BELIEVE) it can achieve." Your brain can achieve anything if you believe it. At first I felt really silly and so I didn't believe, I just kept reading those goals every morning and every night. I remember some of the goals specifically, but I've somehow lost the hard copy. But I can still remember some of them. Most of the goals were designed to experienced what it takes to win; it ignites a burning desire to master the game and win. In losing, we witnessed a model for winning. Winning rewards us with acknowledgment for our training and practice. It teaches us where results come from. It also strengthens our resolve. Win or lose we grow stronger. Where's the problem? The real reward in playing games is aliveness. Think about the times you've played a game so hard you're left panting and drenched in sweat. Didn't you feel really alive? Aren't those some of the best memories of your life? The coaching here is, "Get a game, learn the game, practice the game, and play the game as if your life depended on it...because it does!"

I love to watch ice-skating but just watching an ice show is mildly boring for me. So it's not the skating, it's the competition and the pursuit of mastery that I love. Inherent in my physical game are goals and step-by-step progression in skill and accomplishment. This is an important element. Being acknowledged for progressive stages of skill development also develops self-worth and personal satisfaction. I can play the game of martial arts for my entire life and still grow and expand to new levels of prowess and achievement. I'm not just out to achieve longevity; I'm out to live life fully alive. The element of step-by-step progression also strengthens tenacity. Particularly in studying the martial arts, when you experience these progressive stages of skill development and finally achieve your black belt, you will become a person you wouldn't have normally matured into, no matter how old you got. I can't recognize myself sometimes. Studying the martial arts created new possibilities of who I am and can be. Before my becoming a fifth-degree black belt I couldn't have thought the thoughts I have now or dreamed the dreams I dream now. Also, without the competition, I wouldn't have developed the courage and power to deliver on those dreams.

be completed by the end of that year; I put them in specific terms. The goals went pretty much like this: By December 31, 1982, we will be totally out of debt. That was a very simple goal. Another goal I read to myself was: By December 31, 1982, we will be in our own home.

Still another goal was: By December 31, 1982 I will own a 1983 gold Dodge Caravan that has power steering, power brakes, and air conditioning. I listed all the features I wanted, for not all cars had those features back then, but I knew what I wanted. I even wrote down the color, and I read all those goals to myself, every morning and every night.

At the time, we were not doing very well at keeping up with our monthly rent. Another one of my main goals was that by December 31, 1982 we would own our own home. Remember at the time we were over $80, 000 in debt and had the worst possible credit you could imagine. We were late or behind on every payment and every bill.

The last goal that I read to myself was: I will run the Boston Marathon. I am a runner, and one of my dreams had always been to run the Boston Marathon. I set many other goals, three pages of them. I read them every morning and night. As I continued my vigil I noticed that major things were starting to happen. The very things I was reading to myself every morning and night actually started to form in my life.

This is what happened over that year. My landscape company began to move forward; I did a bid on a large commercial landscape job sometime in the fall of that year. I submit to you that my brain started bringing to me all the things in those goals as I read them every day. I actually started to believe them, and entered them into my brain or, as Napoleon Hill put it, "subconscious mind." As I looked over the year, and as the year progressed, major things continued happening in my life. Anyway, this landscape job came along, and I actually won the bid. Listen to how stupid I was and how careful you need to be about your goals.

This commercial landscape job came through a friend of mine. I met with two gentlemen, who were managing a mobile home park, and they wanted the whole park re-landscaped. I remember thinking, wow, that is a huge park, and they wanted the whole thing redone. To keep it short, I jumped in there, and did the job. I worked really hard, and by the end of the year the job was finished. I sat down in their office, and this is what the gentlemen said to me:

Maybe the source of the decline of aging is a lack of wonder and play in our lives. Kids play games, and look how alive they are. We played late into a summer night, playing kick-the-can...unless it was roller skating season. Do you remember the seasons of our lives when we were young? When kids get an invitation to do something, all you see of them is the door banging behind them as they run out. Kids love competition. They will engage wholeheartedly every time, even though they know they could lose. We played to win, yet win or lose we played again and again. We learned a lot in those games. We learned about ourselves, about being with others, and we developed so much more than we ever knew. We learned skills; we developed our ability to reason, our intention to succeed, our strength, our coordination, our balance, our endurance, our courage, our communication skills, our teamwork, our passion, our love, our full self-expression, and our aliveness. We did this in the safe environment of a game. You could play cowboys and Indians and not really die. Everyone knows that a game is just a game. If you lose, you are not a loser. Just play again and you are a winner.

How does this compare to being grown up? Let's see. If you play war, you can really die and just as bad, you can really kill another person. If you lose at the work game you can lose your home base. The family you have now can decide you are a jerk and really leave you. The "spankings" can kill you. All the games of adulthood have real consequences. These games are serious, grave, even deadly. Wow! Can you come out and play? We are going to have fun killing people today. These aren't games. Games are gone, lost to another time. We invented the Olympic games, but they too, have become serious business. Everyone knows the world of a child wasn't the real world. Being an adult is real. Maybe we have it backwards. Maybe we just forgot

"Okay, Mike, you did an incredible job. We have 1.2 million dollars in our budget; how much money do you want to make?"

And I sat there and I said, "Well, ummm, how about $10,000?"

The man rolled his eyes and said again, "Mike, we have 1.2 million dollars in the account; how much money do you want to make?"

I said, "Okay, how about $20,000?" He rolled his eyes again, and finally he said, "How much money do you want to make?"

I was really bold and said, "$35,000." The guy just looked at me and shook his head.

"That is how much you want to make, not a hundred thousand, not two hundred thousand, you really want to make $35,000?"

I said "Yeah, that would be great." He wrote me out a check for $35,000.

Well, while doing the landscaping job I had already made a whole bunch of money and I had paid down all my bills. We got to the end of the year and were actually 100% debt free. Remember my first goal? I had paid off all my bills; every single one of those debts was paid off. At the end of that year we took that $35,000 or part of it and put it as a down payment on our house. We looked over all the goals we had set. We were 100% out of debt.

We moved into our beautiful new home on December 14, 1982.

At the time I didn't feel like buying a new car, so we didn't own the van, and I hadn't done any marathon training so I hadn't run the Boston Marathon. End of story, right?

Not quite but it is pretty amazing that in one year's time I was able to achieve so many of my goals, just by reading those goals to myself every morning and every night and doing what my brain instructed. If you had told me at the beginning of the year that we would clear off over $80,000 dollars of debt and be in our own home, I would have laughed in your face. It was not even a remote possibility.

Anyway, that is not the end of the story. Probably about five years later we were unpacking some boxes. We had moved, and I was unpacking, and I found that old list of goals I had read to myself, and I started going through it. Here's the miracle and power of the mind. I had actually accomplished every single one of those goals. Even though I had quit reading them to myself, they had all eventually manifested that it's all a game — really. Maybe our play as children was supposed to prepare us for the game of life. Wasn't playing natural? Look at children all over the world. They will turn anything into a game. They experience aliveness and vitality playing their games. We might learn something from looking back at that. Maybe life is a circle and winning is returning to the beginning.

When I got married and had children, as with everyone else, my life became about making a living, creating a home, and raising the children. It was like I'd lost the part of myself that loved playing games. It was when I got divorced that I recovered that part of myself. I married at the ripe old age of 21. By 28, I was divorced. The best part of that first marriage was the gift of my son, Tony, and also that I learned to speak Spanish fluently. I had married a Cuban. We brought my husband's parents to the United States from Cuba. They lived with us and my mother-in-law didn't speak a word of English. Well, all right, she could say "Thank you." Even though my father-in-law spoke perfect English, when they were all together they spoke Spanish. My husband used to say I couldn't stand to not be included in the conversation so I learned to speak Spanish too. Hearing it every day was an empowering environment to learn to speak fluently and have a good accent. I studied it for only two quarters in college and learned to speak, read, write and understand as if I had lived in a Latin American country. People even tell me that I have a Cuban accent in my Spanish.

At 28 years old I was a single mom. I did pretty well at it. My ex-husband was bitter for many years and did not participate in his son's life, financially or otherwise. When Tony was in his late teens his father mellowed out, began paying for his son's college and even acknowledged me for having done a good job with our son. My son, Tony, themselves. Here is the rest of the story. After our move I decided to go out and look for a car, and I bought a used car, and guess what? The car was a 1984 gold Dodge Caravan, and I never even thought about it. The brain is so powerful that every single goal, even that car down to the very color and features was ours. How in the world could I move to another city, go out and look at vans, without even thinking about that old goal, and find this van that was a perfect to fit for our needs and our budget, had the power steering, air conditioning, everything I had stated in my goals? Even the color was the same as stated in my goals and I did not even remember that goal consciously.

You may be wondering about the Boston Marathon. Did I run it? In 1983, I started out and trained, and trained, and trained, and at the time you had to qualify for the Boston Marathon, a 26.2 mile race. It was almost impossible; you had to run a sub 2:50 marathon. (That is less than seven minutes per mile for the whole marathon; 26.2 miles). I trained by running a few miles a day, and naively went out and ran a really hard marathon in the middle of the mountains, thinking that I was just going to go out and qualify for the Boston Marathon at my very first marathon with a little training. I hit the wall at the 21 mile mark, my very first marathon and I hit the wall. I died. I couldn't move my legs, it was my worst nightmare. If a steamroller or big gorilla had been chasing me I couldn't have run any faster. I didn't qualify at that first race, but I almost did. I ran a three hour marathon for that race, which is unheard of for a first marathon. With the very next effort in a certified race in 1983 I qualified for the Boston Marathon. On April 12, 1984 I ran the Boston Marathon. What I am telling you is that my brain brought me every single thing that was stated in my goals. Even though it was years after, and I had quit running. I had quit reading those goals. It wasn't until I found those old goals in that box that I even figured out why I was so driven to run the Boston Marathon. My brain brought me all those things into my life, long after my conscious mind had forgotten about them. My message to you is that the brain is the most powerful thing on the face of the planet. Anything you want to achieve or accomplish in your life is possible through the power of the brain. The first thing you have to do in your youthing journey is to master this very first chapter and you will know why I have chosen was athletic too. One of my joys in life was supporting him at his track meets when he was in high school. In Kindergarten, when Tony was just five years old, he was across the gym and back before the other kids reached the first wall. Becoming a great high school track athlete was not a surprise. I puff up and like to think he got the athletic gene from his mother. During that era of raising a son on my own, I engaged in something I had always dreamed about…riding horses. Again, I didn't just ride; I rode to master the art. I enrolled in the well-known Vogt Riding Academy in Atlanta, Georgia. The Vogts were the ones who rescued the Lippizaner horses from the Russians, who had kidnapped them after World War II. The Vogts imported some Lippizaner mares and stallions into the United States and taught the art of dressage as well as hunt seat equitation. The measure of my progress was of course competition. I focused on show ring hunter-jumper. I loved horse shows and worked my way up from beginner to international circuit competition. I bought my own horse and the best of gear and garb. When I wasn't keeping score for Tony at his decathlon track meets, he was helping me at my horse shows. We raised each other. He was that kind of kid.

My work had taken an important turn at this point in my life. I didn't know at the time that it was important to my development. I suggest that into everyone's life come opportunities, "cubic centimeters of chance" if you will, which, if taken, will profoundly alter the course of that life. The only difference between an ordinary person and an extraordinary person is that the person, who takes the opportunity, becomes extraordinary in the taking of it. One of those opportunities came in my work. Going to work at the same place every day for a good salary was comfortable and secure. In the midst of that security and workability, I got the offer of an interesting new opportunity. The this as the starting place. It is because you have to start by mastering the power of your own brain. Don't let it master you.

Thoughts are powerful things and when you start feeding your brain the thoughts that are necessary for your life to transform, your youthing will truly begin. Remember, I started my journey by reading my goals morning and night. Then I visualized each goal and kept it in front of my brain. Things started showing up in my life.

What do think might happen if you read a simple phrase to yourself every morning and night? You can make it up. It is your journey. Start with something simple like "I am getting younger every day." Try keeping that in front of your mind, morning and night and then see what starts showing up in your life. Remember to listen to everything your mind might bring to you and start doing the things that will bring your goals into existence.

Currently one of the goals that I read to myself morning and night is to look, feel, perform, and run like a 35-year-old. Remember, I am 57. Just the other day I was having lunch with a close friend and one of his associates, whom I did not know. Out of the blue, my friend asked his associate how old he thought I was. The man looked at me for a few moments, sized me up and said, you guessed it. "Thirty-five"! You notice it wasn't 32 or 36, it was 35! There are no coincidences in life. The brain is a powerful, powerful mechanism that is at your command whatever you can get your mind to conceive and believe it can achieve! Notice that there is no "step by step" here. You can have goals related to your weight, your personality, your time, getting rid of anger, what you want to own, what you want work to be like, your family members, your wealth, your religion, virtually anything at all. You are a powerful person. Start on your youthing journey today! Reread this chapter and put your computer (brain) to work. Your brain will deliver. Conceive and believe. Read on.

job offer was training to repair microscopes for a large hospital supply company. Until this time I was working as a certified medical technologist. If I took this job I would be the first woman ever trained at some of the factory schools. I would have to generate service contracts, manage my own accounts, and travel the entire state. The salary was a lot more than what I had been making, and the commissions, as I built the territory, could be great. The real and lasting benefit though, was not the money, but my self-development.

Taking that job was life altering in many ways. First, it taught me independence and self-generation. It developed my courage. Every day I got up and created the work or it disappeared. I sold my own accounts and then delivered. I trained at four factory schools and was almost always the first woman they ever trained. I had to deal with prejudice and then alter their reality with who I was being, rather than have them prevail in their preconceived ideas of what a woman could do. I was trained and worked on $100,000 research microscopes and photo-micrographic equipment. Optical alignments were incredibly tedious and required the finest of attention to detail. I think the work is well suited to the nature of a woman. Once I was called in to align the heads on a seven-station conference microscope. No one had been able to align it, not even the factory representatives. It was totally useless to the hospital without being aligned. It took me 13 hours straight to align it. The doctor told me to bill every hour I spent. I soon became the expert in the state in the tedious optical repairs and alignments. I think the only difference between those who couldn't do it and me, was that I kept going through the dead ends and the frustration. I didn't give up. That reminds me of what a priest once told me about how to have a marriage last forever. He said, "Don't get divorced." That

CHAPTER 2

THE WORD

You Are Your Word

In the New Testament in John, Chapter 1, Verse 1, it says "In the beginning was the Word, and the Word was with God and the Word was God." I have always marveled about the fact that life is made up of words. Think about the power of words. You realize that wars begin and end with words. Nations have risen and fallen by the power of the "Word." People are employed, unemployed and work every day by the power of the "Word." Much of your life is a result of words. Think for a moment when you were a child, and whether you had parents or grandparents or uncles or aunts, foster parents or adopted parents; but whatever your growing up situation was, think about who you are today in regard to that relationship. What you will discover is that whatever you were told (words) when you were a little child in some way, formed your life, many of the things you do, many of the relationships that you have with people, with work, with religion, with family, and even with money.

job taught me perseverance and attention to detail. Those characteristics are elemental in being fully and passionately alive and thus important if you want access to a perpetual fountain of youth.

Servicing optical equipment taught me to do complete and thorough work. I had to notice detail; if I didn't, I would have to drive back across the state fix it. I managed every aspect from selling the work, doing the work, billing the work, to traveling, and billing expenses. It taught me so much. It also paid me well so I could afford another opportunity that then appeared in my life.

That opportunity was a seminar called the EST Training. It was the precursor to the Landmark Education course, The Forum. That was definitely a pivotal point in my life and one that has continued to impact my self-expression, personal power, and extraordinary accomplishments for the past 27 years of my life. I still enroll in every new developmental course they invent. When something works, I use it. When it continues to work, I continue to use it. Feel free to borrow that bit of philosophy.

The Landmark Forum is an educational course that supports people in being the best they can be, having full self-expression, and in having relationships work and be fulfilling, especially with family. That was the hook for me. I'd been divorced for ten years by then and my relationships with men were all the same-they used and abused; either they left or after tolerating way too much, I left, and I suffered. Given that I was the one common element in all those relationships, I wondered if it might have something to do with me. The administrators of the Forum promised me that the situations I'd been tolerating and putting up with would clear up in the process of life after I completed the course. I jumped on it. Doing the Forum was a line of demarcation in my life. The next relationship with a

This chapter is to help you understand that words are powerful instruments. It will also help you realize that life pretty much is what you say. Words are what you use to tell yourself (your brain) what you look like, what you feel like, what you want, what you are going to do during the day and what time you are going to do things, and words pretty much determine whether you are going to be old or young. What we are going to convince you of in this chapter is that you are your word. We will have you understand that you need to be very careful about what you say, because words are powerful things and they actually shape your life. What we want you to do, as an exercise for this chapter, is to think about what you say every day, particularly about youthing. We also want you to look at what baggage you are carrying around. The excess baggage is a result of something that was said to you when you were a child, maybe as young as when you were five years old. And those words we are calling "baggage" actually determine who you are today. That literally means a five-year-old is running your life and that five year old is you. As you identify those negative things, we are going to try and get you to give new meaning to your life. You can alter your life, alter your physical well-being, and even alter your aging and youthing through what you say. Youthing is about words!

This would be a good place to start taking notes. Let's begin with an exploration, a very quick exploration of what you tell yourself about youthing or aging. What you tell yourself about your physical well-being, what you tell yourself about your emotional well-being and what you tell yourself about your financial well-being. We believe all these things are connected to the youthing process, and even who you are to your family, friends and coworkers is directly correlated with what you say. What we want you to do is take a sheet of paper, and in a very honest and direct way try to remember what you say about these things each day. The first thing we'll look at is what you said today about yourself. Try to remember any conversation you have recently had with somebody about you, whether formal or casual, what did you say about yourself? OK, what did you come up with? If you didn't say anything directly about yourself try to notice what you really think about yourself. Come on, you are an expert at this. What is it that you feel? Be honest. Did you say or do you feel that you are attractive? How man turned out to be my husband to whom I have been married for 20 years. I don't think I ever would have been successful in an intimate relationship with a man, had I not done the Forum.

The Forum was an arena in which I completed the past and continue to complete the past of who I thought myself to be. It was a past given to me by the events of my life up to that point and by how I was raised. It was a past, I realized, that limited what I could think, do, and be in life. In other words, in doing the course, I reinvented myself as a creative act, not as a reaction to anything. The other major difference doing the Forum made in my life was having a shift in the way in which I saw the opportunity of life. I got the point that I could make a difference. I previously thought of myself as an ordinary person. Einstein and Schweitzer were examples of extraordinary people. I knew I was not extraordinary. What I experienced in the Forum was that extraordinary people are made, not born. They are made by engaging in extraordinary goals and commitments. I learned that all an ordinary person does to become extraordinary is choose to make a difference in a domain that "lights you up," create a goal in that domain, and take action consistently over time to fulfill that goal. You engage in that endeavor with passion and perseverance. Extraordinariness is the natural fallout when you live to make a difference in the quality of life, not just for yourself, but for others too.

After I completed the Landmark Forum, I accepted an invitation to enroll in a program that was designed to train introduction leaders who would lead seminars introducing people to the opportunity of doing the course. Not only did I not get paid to do that, I paid tuition to do it. As part of that training, I volunteered about 15 hours a week in order to develop myself. The training was six months old do you feel? How young do you feel? Are you depressed? Are you happy? Do you express something positive or something negative about yourself and about life? This is the first thing to look at.

I just heard my son talking to his girlfriend on the phone and here in essence is what he said, "Every one at school thinks that I am stuck up. They don't really even know me. But I can tell they really think that I am proud. But, I'm not. I am really kind of shy and they just think because I don't talk to them, that I am stuck-up." The first question you have to ask yourself is "How does he know what they are thinking?" Who told him what they think? What is his self-image and what is he telling himself, about himself? Even though his conversation isn't about youthing, it is about "self" and will directly affect his behavior. Can you see that many times what you say about yourself and others is made-up? It isn't true and if it is negative, it will probably impact your life negatively. We are going to try to alter that conversation and make it a positive one.

OK, now let's take a look at your physical well-being. What do you say or how do you feel about that? Are you sick all the time? How often do you have a headache? Are you healthy? Do you exercise? Do you smoke? Do you drink? Do you eat too much beef? Are you fat? Are you skinny? Are you self-conscious about any of those other questions? Do you understand that your physical well-being can be directly related to what you say to yourself and to other people about your physical being? Go back and try to write all those things down. Write down everything you can think of.

Take a quick look to see what you say or how you feel about yourself. Once again look and see what you say, is it positive or is it negative? A little bit later in this chapter we are going to analyze all this in relation to the things you can do to help you understand that you can change your conversation and you can also change all the things that are negative. You can change what you say about yourself and how you feel about yourself. You can change your physical well-being and what you say about that is actually how it turns out. Look at your emotional well-being, and also your financial well-being. All these things have something to do with how you feel and what you think about yourself and life. You are what you think and the aging process is a result of what you constantly tell yourself. Words govern your life.

long with three full weekend trainings somewhere in the country and I paid my own travel expenses. With the entire price I paid, in time as well as money, if I spent a thousand times more it would not represent the value I got from that training.

When I was introduced to the Landmark Advanced course, I simply registered for the course. My participation in the advanced course was the source of everything I have accomplished in the martial arts. Until that time I was an ordinary older adult who I'm sure looked pretty ridiculous struggling to keep up with the teenagers in my Taekwondo classes. In the advanced course, I recognized that I always had the power to create my life; I had just created suffering at the hands of others, or being a victim. There is what happens to us, and what we do with what happens to us, is where we have a choice in the matter. In that existential moment in time, I chose to use the divine power of my word to create joy and having a life that I loved. In that course we were given the opportunity to declare before the entire group who we were inventing ourselves to be. It took three days in the course just to rehabilitate us in having the power of our own word and listening to others as having the power of their word. It was an extraordinary space in which to declare who you were and what your life was about. In pondering that rare choice, I remembered the aliveness and passion I had always felt when I was playing sports games. I chose. I declared that the possibility I invented, for myself and others, was the possibility of being the spirit of champion. I didn't choose based on what I thought was possible for a person of my age, then 46. I chose from only what "lit me up," from what ignited my passion. It certainly was not a "reasonable" choice. After taking a stand to be the spirit of champion in the world, extraordinary ways of being showed up. I trained more because it was who I was. It was a natural self-expression.

Let's take a quick look at your emotional well-being. What do you say about that? Are you happy, do you smile a lot and feel good inside when you talk to people or are you depressed? How do you portray yourself? Do you see yourself as an emotionally positive person with lots of good solid relationships, and as emotionally well adjusted, or are you depressed? How about your other relationships? Are you dissatisfied with your emotional well-being? Do you have low self-esteem?

Finally, we are going to take a look at financial well-being. What do you say about money? Are you rich or are you poor? Are you totally in debt or totally out of debt? Do you have lots of credit card bills to struggle with every month? Are you constantly worrying about money? What do you say to other people about your financial situation? Are you positive about it? Do have a good job and make lots of money? Do you always spend less than you make? What are your most valuable assets? Do you own any real estate? Do you have other assets? Do you own your own company? Do you invest in real estate? Stock market? Bonds?

As you look at this piece of paper there should be lots of things you have come up with about your every day life. It is time to evaluate what you say about your situation, and what you say about your life. What we are going to do as a part of this process is look at the power of words and look at the power of changing those words and what you say about your situation, yourself and your environment, and most important of all, what you say about your physical appearance and well-being. In other words, what do you say about aging or youthing. What we really need to look at and evaluate is the actual ability to tell yourself that you are not getting older, but getting younger. This is not just a simple statement. All the things we mentioned influence your youthing and your appearance and it is up to you how you relate to them and then how you interact with other people in relation to them.

We believe that all these things are related to youthing. If you have low self-esteem, if you don't get any physical exercise, if you are depressed all the time, if you now face financial problems and worries, if you are constantly down on yourself and talking badly about yourself, we believe that you will age exponentially. What we want to achieve in this chapter is to get you thinking consciously about the things you say and how you feel every single day about yourself and youthing. Science

I began accomplishing way beyond my experience. I jumped several belt levels at testing. I began winning at international competitions against black belts, when my level was the intermediate blue belt. Twenty years later I still live in the power of that stand. I am a 15-time national gold medalist and a five-time world champion. In the year 2000, in the 18-and-over division, I was fourth in full-contact sparring. I missed qualifying for the U.S. Olympic team trials by one point. In 2004, I was inducted into the U.S. Martial Arts Hall of Fame. My award was for International Female Competitor of the Year. It certainly wasn't because I was the top competitor. I received that award because what I had done represented the true spirit of the martial arts. It was an "impossible dream," yet I engaged in it as if it were possible. I achieved more than any other person my age had ever achieved. I didn't make it to the Olympics. Sometimes when you shoot for the moon, you hit the fence, but it's better than shooting for the fence and hitting the ground! Now there is talk of including a competition in the techniques or forms of Taekwondo in the Olympics in 2008 and my Olympic dream has been re-kindled. I am the current U.S. National gold medalist in forms or technique. I'm thankful to have my Olympic goal again, but my real goal is to be a model and example that being 65 years old can still include being a world-class athlete if I choose to be. I intend to model agelessness. The real accomplishment is in the extraordinary vitality, strength, and passion I still enjoy. The body is not the limit. The only limit is the one we imagine to be true. It's the limit that has the agreement of a whole culture. The "agreement" is that you will, of course, without a choice in the matter, decline as you age. I am here to tell you that I am 63 and I have not hit the wall of that "absolute truth." Every year I am stronger. Every year I am more flexible. The only decline has shown us that an optimist, somebody who has a positive view on life actually lives longer and ages more slowly than other people. There is much evidence of the truth of that. That is a revelation you should look at. Most people who live into their 80's and beyond have been found to be happy, optimistic people, with a positive view of life. We also believe that it is possible for anyone to change their life and their view of life. When optimism becomes imbedded in their existence we call that a transformation. We believe in you, that no matter what environment you came from, no matter what your current circumstances are, that all these things can and must transform. We like the word *transformation* it's much more powerful than the word *change*. You can create a transformation in your life that will create and generate well-being in all areas of your life. Youthing is transformational.

As we started out on this journey we talked about the power of your brain and shared with you that whatever you want, if you keep it in front of your brain, your brain will gradually or eventually manifest that thing into your life. The book *One Minute Millionaire* by Robert Allen and Mark Victor Hansen has many great things in it and we highly recommend that you read it. (Go to our website www.secretsofthefountainofyouth.com if you want to order this book.) One of the ideas that the authors suggests is that everyone "manifests." This means whatever you tell yourself will eventually manifest itself in your life, and if you keep telling yourself the right things eventually you'll have a life you love and everything you have always wanted.

Sometimes you run into people or read in the newspaper or a magazine about older people who start exercising or discover hidden talents in some specialized sport. A good example is Mary Louise Zeller, the coauthor of this book, and her rise to stardom in Taekwondo, the Korean martial art. Read her story in this book if you haven't already done so. As people start exercising an amazing thing happens, they find that their bodies actually can respond to exercise. They start feeling good about themselves. Guess what people are doing? They are all youthing, they are actually getting younger, not older. If you are in better shape this week than you were last week, what has really happened is that you have gotten a week older in time measurement, but your body has gotten perhaps several weeks, months or years younger. You will begin

I experienced is directly proportionate to the amount of work I am willing to do. I think outside of the box of what is generally accepted as normal for my age.

To expand the quality of your life, one of the most important actions you might take is to let the value I derived from the Landmark Forum inspire you to take the course yourself. You will take your own issues to the course, maybe issues you don't even realize you have, and you will go through a major transformation in your experience of life, your relationships, and your self-expression into life. I highly recommend it. You can check our website for more information on Landmark Education. One of the by-products of doing their course is being more aware of life and all of its experiences and thus access being more fully alive. That is definitely one of my secrets to the fountain of youth.

It was soon after my participation in the Landmark Forum that I met and married my husband, Ron Zeller. The Landmark Forum empowered me to complete the painful issues from my past that had up to then defined who I was being in life. I didn't know that at the time. I thought that was just the way I was; that was my own particular personality. When those issues got complete-and, to tell the truth, they have continued to complete themselves at deeper and deeper levels throughout the years since I did that course-I was left with the ability to generate a whole new future. One of the most powerful distinctions I got from that course was that I had always generated life exactly the way it had turned out. I had just been unconsciously generating life as a reaction to what happened to me in the past. In the Landmark Forum you realize that experientially. When you see your own reactions and self-defeating or limiting patterns in life, they lose their power over you and fall away in the process of your living your life. After to understand the power of words. The process of youthing is really possible and you can actually begin to grow younger every single day.

If you do nothing else but tell yourself every morning when you wake up and every night before you go to sleep, "I am getting younger every day," it will actually start to happen.

Through your own words, you are telling yourself what you are or who you are. There are many things you could begin telling yourself. There are also many goals and many different sayings you could use to improve your well-being physically, emotionally, spiritually, and financially, but the most important thing you can do is get started. You must start out simply.

Please, please, please, please, please start today. In the previous chapter, we mentioned the power of the brain, and its ability to bring you anything you want in life. It can bring you health, it can bring you wealth, and it can even bring you youth. What happens when you keep something in front of the brain is that it starts to figure out how to make that happen. Your brain is so powerful that it actually figures out how to bring you those things you want. It alters the universe and things start to show up in your life. You don't have to run out and look for them; they come to you and it's done entirely by words. Words are powerful things.

Many years ago, I decided to program my brain differently, to tell my brain "I choose to get younger every day." Strange things started showing up in my life. Some of the things that happened to me were unbelievable. When you tell your brain every single day "I am getting younger every day," some amazing things can happen in a short period of time. The brain starts believing it and things will start showing up in your life. Because your brain is so powerful, changes will begin to manifest. Do not take lightly the power of thoughts or words in any form they take. This knowledge alone will begin transforming your world. The brain will begin doing that as soon as you start making the statement to yourself, "I am getting younger every day." Start this action immediately. What have you got to lose but the illusion of the decline of aging? Say it to yourself every morning when you arise, at least three times a day and every evening before you go to sleep. Sometimes the simplest things are the most powerful.

the Forum it seemed that I could now own my own power to invent a life I loved. That aspect continues to open up more powerfully year after year. I can honestly say I have a life that I totally love. It continues to grow and expand and I love it more and more.

My husband, Ron Zeller, has been for 30 years and still is a leader of the Landmark Education courses. Marrying him has been one of my most valued results of participation in the Landmark Forum. I say it was one of my results, because I wrote on my Forum application that the result I wanted to produce out of my participation was a marriage relationship that fulfilled on the promise of "oneness" between a man and a woman. That was certainly one of the most important issues in my life that I said I wanted to alter as a result of my participation in that course. It certainly did alter. After I completed that three-day course, the particular kind of man — I'll call the type user/abuser — that always seemed to find his way into my life, never showed up again. My "magnet" had changed.

At this point I should also mention another course that expanded my experience and expression in life and that is the Tony Robbins course. To tell you the truth, the only thing I remember from the entire three-day course and the only thing that made a lasting difference for me was the fire walk. Yup, the fire walk is real. It's not a trick. I walked about 25 feet, barefoot, on live hot flaming coals and did not get burned. Of course it took about six hours of specific mental preparation, so don't go out to the backyard and try to do that on your own. When I got to the other side of that flaming path it was an experience I'll never get over. It was the "twilight zone." The major impact of that experience for me was to know without a doubt that we truly have no idea of what we can do. It also clearly demonstrates to you the power of your mind to generate reality. You can

The first thing that usually happens is that the things you already know, and yet aren't acting upon, begin to surface. Once you begin, your brain simply starts talking to you and telling you the things you need to do. Part of the process is to listen to what your brain says. Listen intently. One of the keys to growing younger, you already know, is exercise and diet. I promise if you keep telling yourself, "I am getting younger every day," your brain will tell you how to change your diet and to start exercising. Guess what? One of the key elements to feeling good about yourself is — *exercise*! The body and the brain both know that if you will exercise you will immediately start getting younger. Even if you are out-of-shape, *begin*. Begin from right where you are. You don't have to lose weight to begin. It's in the exercising that you begin to lose weight. As soon as you begin to exercise, the brain will swing the whole body into a building mode. It will be yelling the command, "We need more muscle! Build! Build! Grab the fat stores! We need more energy!" Intense movement and demand on muscles and endurance signal to your body to begin youthing once more and that you are not slowing for your exit. Exercise is life!

As things begin to change, your body does actually become younger. Day by day and almost every day, you get younger. Your brain may tell you don't eat that thing you are going to eat. How many times have you eaten something, and you know you are not supposed to. Almost every minute of every day there will be a choice to make that will give you one future or another. What you have to understand is that truly one of the most elemental factors contributing to the decline associated with aging is what you eat. "You are what you eat." One of the best things you can do is to start listening to your brain and to start eating those things you already know are good for you. You need to start eliminating the things you know are not good for your body. See Chapters 6 and 7.

You will also begin to see other things in your life that you need to adjust or rearrange. We talked about emotional well-being and what you say about yourself. Sometimes you might say to yourself that you are depressed or you are angry or whatever your body is making up at the time. You can be depressed and you can be angry. But, what you have to do is take charge of your life, all aspects of your life. In order to move beyond your own abilities you must consider yourself being relate this to what I call the myth of the decline of aging. I think we decline as we age because we believe that is what happens to human beings as they grow older. Doesn't everybody know that? If someone doesn't decline as rapidly as everyone else he must just have good genes. Yes, that explains it. If that were true it would be bad news. The good news is that the decline of aging is a choice. At the age of 63, I am vitally well, stronger, and more fit than most people half my age. I still compete in a full-contact sport, requiring strength, speed, agility, endurance, and balance. I beat people half my age regularly. I am still the current national gold medalist in technique in the 18-years-and-older division and have been for the last 15 years. I'm the most over 18, to have ever done that. It's not an accident of genes. It's been a project and I can share with you what I have discovered that enables me to continue to play and live out my dreams for the rest of my time here on earth, no matter how old my body gets.

After being a single parent for 10 years, choosing to marry my husband, Ron Zeller, was yet another example of leaping at the opportunity life brought, although it took me three years after I met him to "leap." Marriage is always a risk, even though the way many of us deal with that risk is to not think about it too deeply or to avoid it completely. The relationship of marriage, I say, is the most complex and difficult of all relationships. With Ron's training in human behavior and communication, I knew that in marrying him I would reduce the risk. I can still count on Ron to transform every breakdown into a breakthrough. When he makes a commitment, he is true to his word–always. When you exercise your power to always keep your word, you actually strengthen your power to create all your desires into reality.

Ron is the first man I think I have ever fully trusted. He has been and always will be committed to my highest and financially well off enough to have everything you need in life. It might be exercise equipment or good food, or it might be, like in my case, changing your diet and your supplements. You are what you eat isn't just about the "what not to eat," but also about the "what to eat." Over the years I have gleaned, studied, and learned many things about nutrition, as has my coauthor, Mary Louise. We present them to you in this book. The point is that if you will consistently tell your brain, "I am getting younger every day," it will deliver all the things you need to be on your lifelong journey to youthing. Your life will transform.

You can begin to understand the power of words and how to present them to the brain in such a fashion that words will begin to have extra meaning in your life. Words are not cheap! It will help you to understand that you can use words to begin your journey into the world of being younger. You can move forward now to a new level of awareness and begin to transform your life into a life you love, a life of getting younger by the words you choose to feed your powerful computer — your brain. Remember your youthing journey begins and ends with words.

best, so I value and trust his coaching. Can you imagine listening to your spouse and allowing him to contribute to you without feeling like you are being dominated or having your freedom to choose being diminished? I think a good marriage is important. Some of us choose a new outfit to wear with more consideration than we choose a life partner. I have a dear friend who had a good and successful husband. He thought her life should be only about him and she should serve his every wish and she did quite willingly for 25 years. If you knew them you would say they had a good marriage. They really did love each other. They just didn't know how to communicate at the level of saying what was most important. She sort of gave up her own growth and expression and only served her family. I serve my family too, probably to a fault, but I also am passionately committed to my own growth, learning, and self-expression. Even though my friend willingly gave up having any attention to her own needs, she was angry underneath it all. It boiled over in her speaking from time to time. When my friend's children were grown, she got cancer and died quickly. Out of my conversations with her I am sure that she chose to leave that way. She loved him too much to leave him in this life, so she left the only honorable way she could. I'm not saying you should believe me or that I believe myself, but I do entertain the possibility that we may have the power to choose to leave this earth by choosing to die.

My husband has generated his life to make a difference in humanity; to expand all its possibilities. This is one of the secrets of the fountain of youth in case you missed it. Consider it underlined. Ron lives with incredible daily commitment. He's my model of living an extraordinary life. He's 73 years old and looks about 55. He has the vitality and strength of a 30-year-old. He still takes on very big goals. Retirement is an insane idea to him. Why would you

CHAPTER 3

DECLARATIONS

Breathing Life Into Your Words

This is one of the most exciting chapters. The reason is very simple. In this chapter we will provide effective access to the most powerful tool in the world, your brain. This chapter is meant to empower you not just in goal setting, but in creating powerful declarations, proclaiming them to the world, and transforming your life through the power of words and the power of the brain. Your words can consistently and literally move mountains. The secret to having a youthful life is not just exercise and diet, but is also a simple belief that youthing is not just a linear process, but rather a profound frame of mind takes responsibility for what you tell yourself and other people every day. You can literally dare to create your own great self-expression.

The process of what you tell yourself and others has been called "making declarations." In this chapter we will present more specific detail on empowering your declarations to become reality. We have been involved in some incredible training, and we highly suggest you too take on this training as an integral secret of the fountain of youth.

quit creating and producing, especially when you, at last, have enough experience to really make a difference? Our tombstones will read, "Burned Out in a Blaze of Glory." At the beginning of every year Ron sets many new goals. The difference between him and many well-meaning New Year's goal setters is that he actually does everything on his list...and I mean everything. If it's on his list, it's done. To celebrate his 65th birthday he ran the Wasatch 100, a 100-mile-race up and down 11,000-foot mountains from Ogden to Sundance, Utah. The course goes up and down seven mountains; from 5,000 feet up to 11,000 feet, back down to 7,000 feet, back up to 10,000 feet, and so forth. I was his pit crew. The runners stopped for about 15 minutes every 15 miles and took in food and re-hydration drinks. I was there at each station to also wash his feet, rub down his feet and legs, and give him fresh socks and shoes. In the end the foot care paid off. He was one of the few who did not have one blister. Probably 25 percent of the runners dropped out because of severe blisters. When he wanted to quit at 75 miles, I poked food in his mouth, pulled him to his feet, and pushed him out the door saying, "You said 100 miles, not 75." Ron not only finished that grueling race, he finished first in his division. We often only succeed in our impossible dreams because of the support of others, i.e., it's not smart to go it alone.

Ron loves me–really loves me. I don't know that I ever would have had the experience that I am loved if it hadn't been for him. He has been an environment in which I have thrived and grown. It seems the more he empowers me, the more I love him. I laugh when I think about the time I met Ron. He told me that if I was a woman who wanted to get married and have a child, I probably shouldn't go out with him even once. I think one of the reasons he liked me so much was that I looked him in the eye with a straight

A company called Landmark Education provides this extraordinary educational process. They have a series of training courses or workshops that provide a safe and effective curriculum for you to have life-altering breakthroughs and transformations. Words are, therefore, a part of any transformation, breakthrough, educational milestone, or any other evolution of intelligence we are able to have for ourselves in this human existence. Words are the source of power in this life.

When you make declarations in your life that is, declaring your goals, clearly stating that which you wish to obtain, and having life be exactly how you want it to be. Declarations become a powerful tool in transforming your life and launching you on your journey to a younger, more vibrant you. Youthing begins with declarations. So, what do declarations look like? How do you make declarations, and keep them present? How do you get to a point where your brain starts believing your declarations and actually starts transforming your life? Those are the questions we will answer in the next few paragraphs. Pay close attention. Begin to structure your declarations in a way that empowers you. As you form your declarations, you must attach emotion to them. You must believe them yourself. You must convince your brain you will have the things in life you authentically say you want. The way in which you express these declarations will determine if they work.

First of all, your declarations must be powerful; you must be able to state them in such a way they empower you. Weak declarations are just words without real power. Let's begin with some simple things you can do to state your declarations powerfully. Keep the words simple. Declarations are not fix-it techniques. Many times we believe we have to fix our lives; therefore our declarations are intended to be a way to fix something wrong in our lives. The truth is declarations are powerful words spoken into a clearing. A clearing is a state of emptiness. It has nothing attached to it. There is no past. There is no present. There is only a future. Declarations are delivered into this clearing. This means declarations aren't to fix something; they are statements about the way you want life to become rather than saying what you don't want. Declarations are powerful tools if spoken correctly, will alter your life and transform it into a life you love. Your life will be filled with youthing, and the future will be filled with the things you want in life. Declarations create face, and said, "If you are a man who doesn't want to get married or have a child, you shouldn't go out with me even once — so, now what?" We went out. We also got married and had a child. I tell him all the time that I am the one with the strongest intention. One of the things I learned in my perseverance is that some things take time. It took three years for us to choose to get married. Our lives have not been without ups and downs and even tragedy. After two heartbreaking miscarriages, my second son, Adam, was born a year and a half later.

Adam was born when I was 44 years old. He was and continues to be a great joy in our lives. He also taught me things I know I would never have learned any other way. Having children late in life may be another secret of the fountain of youth. We tried to have another child, even though our friends thought we were crazy. I miscarried another two times and never again got pregnant. When Adam was 19 months old we almost lost him too.

Ironically, it was Mother's Day, and we went to visit friends and to see their new house. The neighborhood gathered for a welcoming barbecue. Adam went inside the house with my friend and me as she showed us her new home. They had not yet moved in so the house was empty. Her children were older, about five and seven. Adam toddled around the empty rooms, chasing after the older children. I held his hand going upstairs, and as we walked down the hall, he pulled out of my hand to run after the other children. I was not concerned. The house, being empty, looked safe; there was nothing to climb up on or get in to. As I came into the room after him, I was talking to the other mom and looking around the room. As my eyes reached Adam, I saw him reach out to touch the screen of a ceiling-to-floor window. The screen gave way under his touch and he fell forward out of the window. I took a step a future. For example, "I am growing younger every day." "I have energy at a level beyond imagination." "My body is in perfect condition and my mind is keen." "My face and all the elements of my outer expressions are youthful." Feel free to use these or create your own. Write them and then speak them. Speak them in the morning, some other time during the day and at night, just before going to sleep.

Another way to expand your understanding of declarations is to give you some examples of some powerful declarations people have made with the results they have produced with them. Without knowing it, many great leaders, and people of wealth, fame, and fortune, have made declarations that produced for them the very things in life they spoke into existence.

Many times, people have spoken in public and made declarations that have had a lasting impact on people around them, their own lives, and even the world. Here are just a few examples.

Declarations by famous and infamous people

Johann Wolfgang von Goethe – "Before you can do something, you must first be something."

Mother Teresa – "Spread love everywhere you go: first of all in your own house."

Adolph Hitler (infamous) – "Strength lies not in defense but in attack."

Abraham Lincoln – "All that I am, or hope to be, I owe to my angel mother."

Eleanor Roosevelt – "I believe that anyone can conquer fear by doing the things he fears to do."

Thomas Jefferson – "One man with courage is a majority."

Benjamin Franklin – "Energy and persistence conquer all things."

Sir Winston Churchill – "History will be kind to me for I intend to write it."

History has validated these famous declarations. People's lives, nations, and the world have been powerfully impacted. Your life and your journey to youthing don't just begin with a declaration. This declaration can ongoingly transform you for the rest of your life. A single declaration is really that powerful.

toward the window, even seeing that he was already gone. As I remember that moment, I interpret that I suffered a severe mind shock. The room seemed to turn into a long, dark tunnel. I told my husband I couldn't even scream. He told me later that I was screaming. I didn't remember it at all. I remember pushing past the woman's husband, who was running up the stairs, probably because I was screaming. He grabbed hold of my arm and I struggled until I got free. I heard Ron's voice yelling, "What happened?" I yelled to him that the baby fell from the upstairs window!

As I burst out of the door, I saw Ron kneeling over Adam. Later I found out that he was checking for a pulse and breath. How traumatic that must have been for him. Adam was sprawled face down on the pavement and just as I came out of the door, he raised his head up and began to cry. At that point, my blessing was in knowing that he was alive. Not knowing how badly he was hurt, I knelt beside him cradling and soothing, waiting the eternity until the ambulance arrived. The paramedics arrived and took over, checking and stabilizing. I rode to the hospital in the ambulance with Adam. Ron followed in the car. At one point in the ambulance, Adam's eyes rolled back in his head like he was going unconscious. I called to him to stay right there with me, to look at me. He came right back and remained conscious the entire time thereafter. After his immediate evaluation and treatment in the emergency room, he was taken to X-ray. His father arrived and went into X-ray with him because I was six weeks pregnant. Adam had skull fractures in three places. They said his mandibular joint was fractured, but was the smaller threat. They took him in to do a brain scan. Adam was fortunate. The fractures were simple with no indentations and therefore no damage to his brain. The doctors told me he would have some swelling in his brain from the trauma, but it would resolve itself in about six weeks.

What we want to help you do is to make declarations and to make them in such a way they can literally transform your life. Just take a moment to review all the powerful declarations listed on page 62. What do they all have in common? Notice declarations are spoken publicly. In fact, declarations become greater as the number of people listening to them is greater. The greater the number of listeners, the more likely they are to become a reality.

Declarations are to be stated powerfully. You must choose words that will literally bring the declaration into existence. Once again, look at the examples and notice the words. One of the things should be evident is that they are very simple. A powerful declaration is weakened by an excess of words or with a lot of fluff. Declarations are powerful, to the point, and spoken with authority. Some people in their journey to youthing make a statement something like "I am getting younger not older." A more powerful way to say the same thing is, "I am not getting older. I am getting younger." The reason why this is more powerful, is simply that the wording of the statement first negates the past and then creates the future. An even more powerful statement would be, "I am getting younger every day." The reason why this is the most powerful way to speak your declaration is that you are not trying to fix something. Arguing against the opposite weakens the declaration. You are simply making a statement that creates a new future.

In *One Minute Millionaire* the author suggests six key areas in which to set goals or make new declarations. They are BODY, BRAIN, BEING, TIME, PEOPLE, and MONEY. I have created six such declarations or goals and I write them out every single morning and read them aloud five to six times every day. Somehow these declarations continue to be fulfilled. I believe that "somehow" is the inherently divine power of the brain acting upon what I believe. As you bring yourself to believe those declarations, as you express them with emotion, get excited about them, and read them more and more, with more and more feeling, you can actually speed up the process of fulfillment.

Think through the many simple declarations you could make in relation to your journey to becoming younger. One of the things that makes declarations powerful is to not have too many of them. If you have too many declarations, your mind will become confused about

It was a difficult time in our marriage. Ron and I dealt with Adam's accident in different ways. Ron was intellectual about it. Adam was going to be OK, now let's get over it and get on with it. He threw himself back into his work and got on with it. Ron was traveling with his work all but about eight days a month. When he was home he was on conference calls and preparing for his next assignment. There wasn't a lot of relating going on. He especially didn't want to deal with my emotions about how I was coping with Adam's accident. He was making me wrong for being so emotional about it all and I was making him wrong for being so unfeeling about it all. He handled it by doing more work and I handled it by overprotecting Adam. I felt alone and angry with Ron that he wasn't there for me. Instead of listening to me and just expressing understanding and offering comfort, he tried to "fix" me so I would get over it. He got angry and impatient with me when I was emotional. This is not at all unusual for a husband and wife when they have a child seriously injured or killed. I think the statistics show about a 75 percent or higher divorce rate.

The week Adam came home from the hospital, I miscarried. Ron was out of town. I had a friend come and stay with Adam and I drove myself to the hospital. I had to have a dilation and curettage or D&C, a minor surgical procedure to prevent infection after a miscarriage. Along with the emotional trauma of witnessing the near death of my child and losing an unborn child, my own hormones betrayed me and wrecked more havoc on my already out-of-control emotions. I literally cried most of my waking hours. Taking care of Adam is actually what held me together. Then, in the days following, Adam had a few seizures, which scared me to death. We had a few mad dashes to the hospital. That is not too unusual with the inflammation and swelling in the brain that goes on what you really want. As to the powerful declarations from famous people, there may have been many other declarations but the one that sticks out is the one that is the most powerful. Look over the many declarations you want to make, about what you look like, what you feel like, what you eat, how you train, what you do every day, and choose a statement you see yourself becoming. It will light you up.

Make a declaration that would cover all those areas in one powerful statement. Speak that statement out into the world, to as many people as you can. You can post it where you'll see it all day. You could even demonstrate it as a collage of pictures you cut out of magazines. You can record it, input music quietly into the background, and play it as you work. Watch your life change. No, watch your life transform.

in the weeks following a head injury. The doctor told me to make sure that Adam didn't bump his head. Can you imagine the exhaustion following a toddler around all day making sure he didn't bump his head? I'm so grateful for the ladies from my church. They really live the charity of Christianity. They sent women over for four-hour shifts while I was recovering from my miscarriage. They cleaned my house, cooked for us, and even ironed my husband's shirts.

Adam was 19 months old at the time of his accident and his speech was advanced for his age. After the accident he didn't talk at all for about three months. He dragged his right leg in an awkward gait. I was worried and anxious about him all the time. He had a lot of neurological testing and physical and speech therapy for about six months. It was such a hard time. I don't know how people deal with injuries and illnesses where there is no hope of getting better. I'm pretty sturdy and handled it like a trooper, but my hormones were whacked out after the miscarriage and again I subconsciously harbored anger at Ron for not being there, when in reality all he had done was work to provide for us. I have a tendency to "stuff" my feelings under the surface and carry on like it's no big deal. In the past I just told people what I thought they wanted to hear. I wanted to please. I've had to learn to be aware of my own feelings and communicate them straightforwardly. Even when Ron doesn't want to listen I now keep going and press into it. I really am committed to filling his needs and I had to learn how to make clear requests for what I needed. I have to acknowledge the Landmark Education courses again not only for teaching me how to communicate, but also for making me aware of that domain. I thought I communicated and knew myself when in fact I didn't have a clue what was real communication.

CHAPTER 4

HEALTH AND WELLNESS

You Are As Young As You Feel

"It is part of the cure to wish to be cured" – Seneca said this 2000 years ago.

Before we can think about youthing, we need to address the primary element of health and well-being. The opposite of health or well-being is sickness. There will be many who read this book who do not have 100 percent health or 100 percent wellness. The opposite of health is sickness and many people are chronically ill with many disorders. Things aren't going well for them in the health arena. They may have a disease, or some complications of disease. They may have an advanced disease such as cancer, muscular dystrophy, multiple sclerosis, Alzheimer's or any number of other serious illnesses. What we want to look at is, "Can we restore health?" Can we bring back a level of health and wellness for people that would allow them to reclaim their lives and in the process even reverse the aging process? Certainly part of the "normal" aging process is as we get older, we will have more health problems, more illness, more disease.

I actually had no idea of what I was feeling. It was like I was living only on the surface of life, never looking at the immense swirl underneath. Ron's a really bright guy, and even though he certainly didn't have a model for a good marriage relationship, given his upbringing, he was very willing to listen when I began communicating correctly. When we really communicate, there is no anger. When real communication is present, so is affinity. I certainly didn't have any education in communication in all my years in school. I think there is a desperate need for that kind of education if we are to reverse the high divorce rates in this country. Anyway, we got through it and didn't get divorced. We pulled the statistics a little in the other direction.

At the time of Adam's accident, I was working on a master's degree in psychology and counseling. After his accident and my miscarriage, I was so emotionally distraught I couldn't concentrate on my reading and other work. It took me an hour to read a page. I got to the end of a paragraph and realized I had no idea of what I had just read. I reread the same thing and lost my concentration a few sentences into the paragraph and had to read it again. I couldn't cook a meal and have everything done at the same time, when I had been multi-task oriented before. It took me forever and a day to clean the house. I'd go to get something in another room, get distracted, forget what I was after, and engage in a totally different project. I'd drive down the road and just burst into tears. I started having fearful imaginings. When I washed the dishes, I imagined someone might drive by and shoot in the window at us. If I took Adam to the park in the stroller, I imagined that a car would jump the curb and hit us. What I have described are classic symptoms of PTSD. Post-Traumatic Stress Disorder doesn't just go away. Many people who experience traumatic events keep going into those symptoms until they go off

Wellness is a whole different story. Wellness has to do with many seemingly insignificant health protocols, that when consistently followed will continue to produce health and wellness all the days of your life. We divided this chapter into two categories, health and wellness. Wellness is ridding our bodies of all aches, pains, and malfunctions; wellness is having a body that can always exercise and perform, having a body you love. The whole health and wellness section is delivered from that point of view and in relation to youthing.

You must have health and wellness in order for youthing to take place, because if you are getting younger means you are sick less, if at all, and you are able to keep on raising your overall well-being to an advanced level. First, let's take a look at health and various sicknesses, and what can be done about it.

The level of health and wellness you enjoy is directly related to the function of your immune system and the function of all the systems of the body. The skin is an organ and it functions as a system. You breathe. That is the function of your respiratory system. You digest and utilize food. That is the function of your digestive system. Your blood carrying nutrients and oxygen to every cell and carrying away cell waste is the function of your circulatory system. There are many other functions and influences that contribute to your overall health.

One of the keys to extraordinary health is having a strong immune system and strong immune function. I don't want to get into a long physiological, scientific explanation of the immune system, but let's break it down into something that is very easy for us to understand. Since I don't know if everybody understands the way our immune system functions, I'm going to put it in the simplest of terms and we are going to look at it from that point of view. How do you measure health? One of the ways is by how often you get sick. If you look at any group of people, you will see patterns of health emerging. Many of these health patterns are hereditary. Look at yourself, look at your family, if you are married, look at your spouse or your children. Look at your relatives and you will notice different positive and negative patterns of health and physical function. Everybody has concern for his or her health to some degree. This concern usually grows as we get older. Much of that concern is based on how sick or how well we are and how well our ancestors were, these the deep end. I decided to get some counseling. I went to someone who specialized in Post-Traumatic Stress Disorder. It was a very good choice. I was getting a grip back on my life. The symptoms were receding more each day.

It was during this time of recovery from Adam's accident that I was sitting at the kitchen table one Sunday morning, reading the newspaper. I noticed an ad for a martial arts school of Taekwondo. It said something about improving physical fitness, mental focus, and self-confidence. Well, my self-confidence was shattered. After all, I couldn't keep my own child safe. After four miscarriages and the birth of Adam, I was in the worst physical shape of my entire life. I was 25 pounds overweight and I couldn't even run a block. At that time, I would have described myself as "handicapped" in the area of mental focus. I thought I should try this Taekwondo thing. It certainly looked like it would be more fun than running, aerobics, and weight lifting. When I walked in the door of Master William Kim's Family Taekwondo Training Center in Vallejo, California, I can't imagine what he thought when he found out I was inquiring for myself and not for my child. It probably went something like this, "Well, at least it'll be one more paid tuition." Actually, I'll always be eternally grateful to Master Kim for not treating me like a middle-aged woman who must be nuts for wanting to learn Taekwondo. He treated me like any other student and expected me to do what every other student did. I felt like a beached whale and all the teenage students in the class were agile dolphins. I felt ridiculous when I tried to do the techniques. Master Kim probably thought I was slightly retarded as well. I had him repeat a Korean word at least four times before I could say it myself. When you learn Taekwondo, you learn all the commands in Korean, much like what the French did with ballet. You could study Taekwondo anywhere in are key factors. Much of physicality is hereditary and you might have good or bad health as a result of what you inherited from your parents. You need to look at the history of your health. For example, when did you last visit the doctor's office? What does your family medical history look like? Do you have cancer in your family history, heart disease, high blood pressure, or diabetes? This genetic influence can contribute to your level of health. You could say you are predisposed to be a certain way regarding your health. If you are predisposed to be diabetic, for example, is there any way to nullify that predisposition or to completely block it from ever showing up in your life? One of the secrets is to empower our immune system beyond hereditary predisposition for an overall expansion of health and one of the keys to youthing is to discover how we can do that. What part do all the influential factors play in our health? Does heredity have to affect our health? What part does diet play and what part does exercise play in being able to live a long and healthy life?

Immune function

I'm not a doctor and don't pretend to even be an expert in human physiology. However, I have read a lot about immune function and, being a serious athlete, I have been my own subject of study for years. I don't pretend to understand the whole thing, but what I do understand I'm going to share with you and perhaps you can learn from all my experience. If you are a doctor or other medical professional please excuse this very basic and unprofessional explanation.

Doesn't our immune function determine how well we are able to combat sickness or disease? Does it also determine how well we are able to avoid and combat disease is related to our heredity? If we are predisposed to have a weak immune system, does that always have to determine our level of health? Is there any way to strengthen immune function? These are some of the questions we will try to answer.

Immune function has to do with these little things that run around in your body. You have things that are called oxidants and antioxidants, you have things that are called free radicals, you have things that are called T-cells, you have red cells and white cells, and all of those "things" are in your blood, your respiratory system, your lymph system, and your digestive system. These are the things that together constitute your immune function.

the world and not have to speak the language to follow along in the class. That kind of standardization is what helped Taekwondo do to be the only kicking martial art in the Olympics. In the year 2000, Taekwondo became a full-medal Olympic sport. Having a worldwide governing body, the World Taekwondo Federation, and having the same curriculum taught the same way in 170 member nations around the world, is the other main reasons why Taekwondo has achieved Olympic status.

In the beginning, why I kept going through all the frustration, feelings of inadequacy, and the horrendous muscle soreness, I can't imagine. Another adult told me he tried Taekwondo for about a week. He said it was too much for him. He didn't think an adult could be flexible enough again or have the balance, strength, or endurance that Taekwondo required, much less deal with the muscle stress and strain at a mature age. Taekwondo was just for kids. He quit after a week. He encouraged me to give it up too. I was older than he was. He was certain I wouldn't be able to continue. That might have had something to do with why I kept going. After three days, my muscles were so sore I literally couldn't walk downstairs. Really…I sat on my bum and slid down the stairs. Even when I was taking Ibuprofen, the pain kept me awake at night. I finally solved the sleeping problem by putting muscle-relaxing bath salts in a tub of hot water and falling asleep that way. When the water got cold, I woke up and ran more hot water. I slept this way for a solid two weeks. It's a wonder I didn't drown. Instead I got stronger. If you had told me at that time, though, that I would compete at a world-class level in Taekwondo, a full-contact sport, much less be a fifth degree master instructor, I would have thought you lost your mind.

After six months of training in Taekwondo, my symptoms of post-traumatic stress disorder disappeared

What happens in your immune function? Whatever it is, something allows you to battle disease better when you are younger than when you get older. That should be obvious to you. As you get older it is much more common to get sick and have a variety of malfunctions and diseases. In the flu season we are even issued warnings that the very young and the very old are much more at risk. Let's take a look at the immune system.

One of the miracles of our immune systems is killer T-cells, also called T-cells, which are good things in our bodies. "Killer" doesn't sound good, but they are not killers in a bad sense, but killers in a good sense. They kill off the bad stuff. We'll just call them T-cells. T-cells can be counted and their job is to go around looking for bad cells or anything that would attack the normal function of the body, such as bacteria or viruses. They are the police force or the army in your body that goes around and looks for malfunction or for disease. Once foreign invaders are found, the body rallies the army of T-cells to kill off all bad stuff. Take cancer for example. No matter what age you are, your body produces cancer cells every day. Every human being produces them. What keeps you from getting the disease of cancer is these T-cells hook up with the cancer cells, identify them, and call out the whole army and its arsenal to destroy them. When you receive a diagnosis of cancer your cancer cells have reached a reproductive stage where they are producing and reproducing themselves faster than the T-cells can kill them. Your army is losing the war. You then begin having symptoms. The cancer cells grow to the point they produce a tumor, which literally means, "new growth." In our analogy, the cancer cells have captured some of your territory and built their fort or new base from which to launch a bigger attack on your body. If it's in your bloodstream you know it as leukemia. You then receive the diagnoses of cancer. The T-cells have not identified and killed off the cancer cells at the rate you needed to prevent the diagnosis of the disease of cancer. That is pretty interesting. I didn't understand that before. The fact that all of us produce cancer cells every single day was news to me. What helps us to avoid reaching the point at which we get the diagnosis of the disease is immune function. The function of the T-cell cells is to search out, hook up with a suspect cell, communicate with that cell, and determine if it's an unrecognized or foreign cell. If completely. My counselor told me the normal course of therapy would be about two years. I could expect my symptoms to slowly diminish over that period of time until they completely disappeared. She became so interested in my accelerated progress that she began recommending Taekwondo to all her PTSD patients. She theorized that the cross patterning associated with Taekwondo developed new neuro-pathways in the brain for focus and concentration. In my own experience of teaching for the last 17 years, I also theorize that is why children with attention deficit disorder improve practicing Taekwondo. She also thought that the self-defense aspect rebuilt self-confidence. Experiencing how fragile life is through an accident shatters the illusion of the continuation of life. In other words, we naturally expect to keep living. Accidents show us that we, or those we love, can be snuffed out at any moment-safety is an illusion. Life is dangerous. When we become skilled in martial arts, we become stronger in the face of the risk that life is and literally grow more self-confidence. Studying martial arts is the fertile ground that grows self-confidence. It's not just a marketing slogan. Engaging in a full-contact sport also generates courage. It demands you be fully present. If you're not, you'll have a foot in your face. Pain is sometimes a good teacher. In PTSD, the symptom of lack of focus is the human being's way of not wanting to be present to the danger and pain life can deal us. We don't want to be aware of that. We are trying to avoid the truth of that.

Taekwondo demanded my being present again. It was like someone threw me a life preserver when I was drowning. Taekwondo was bringing me back to life.

Competing in Taekwondo was something to which I had never aspired. It never even occurred as more than a thought flitting through my mind, only to be immediately dismissed as temporary insanity. When Master Kim first it is, they call out the army and kill it. That is why they are called killer T-cells. Over time, if we don't have enough of those T-cells or they are not functioning well enough to communicate with or kill off the invader cell is that cell escapes, multiplies, and creates disease.

If we were able to somehow strengthen immune response function, that is get those T-cells communicating and doing what they are supposed to do, we could defeat any disease or malady. We would always be in a state of good health. But the whole function begins breaking down with age. That breakdown may be triggered by heredity, diet, lack of maintenance and strengthening through exercise, or from years and years of exposure to toxins and poisons from our environment. We may be genetically predisposed to a certain disease, or for these cells to not function optimally. Some new research that has surfaced has shed new light on the subject of cellular communication. I believe that research came out of the state of Texas. It showed there are ways to help those T-cell cells communicate better. This research started over 20 years ago. Some doctors, both medical and research types, started looking into what causes cellular communication. They developed a whole new viewpoint regarding cellular communication. This new viewpoint was a departure from the previously accepted viewpoint of the medical community. Cellular communication is what our immune systems use to prevent disease in our bodies. When immune function breaks down we get a disease. Cancer is an example of what happens as a result of immune dysfunction. The disease of lupus erythematosus is an example of what happens when good cells are destroyed by T-cells. This is called an autoimmune disease. In other words, a T-cell hooks up with a normal cell, misidentifies that cell as a bad cell, and destroys the good cell.

Many of you have heard of the disease of Lupus. Lupus occurs when those T-cells start taking off and killing off all the good cells. Your body literally starts eating itself. That is a simple definition of lupus erythematosus. Rheumatoid arthritis is also an autoimmune disease. Autoimmune disease can be explained as the T-cells acting erratically and erroneously. Normal cellular communication is not taking place. The T-cells start killing off the good stuff instead of the bad stuff. In dealing with disease, opposite dysfunctions equally produce disease. On the one hand, the T-cells aren't identifying disease-producing cells, invited me to a tournament, I thought he wanted me to help out...to drive for some of the kids, to make sandwiches, to cheer lead...not compete! He said, "No. You will compete." It sounded like the government's invitation for me to pay taxes. I just said, "Yes, sir" and registered for the tournament. When it came time to really compete, I was so nervous I could hardly breathe. I did it anyway. When it was over I realized I hadn't felt that alive for so long I couldn't even remember. I don't think I had that feeling since I was in high school playing on the basketball team. Before every game the butterflies in my stomach were at the same time almost unbearable and yet were so stimulating it was nearly addicting.

After that first tournament, I was hooked. All I wanted was to learn more and get better. I wanted to win. It felt great. It was at the time I wasn't getting what I needed from Ron. Taekwondo made me feel better. It comforted me. It was there for me. It gave me what I needed. I got very good at Taekwondo. I lost weight. It was an injection of aliveness and power. I had a friend in my Taekwondo class, a young woman in her late twenties. She came every day like I did. Her husband asked her accusingly one night, "Are you having an affair with your Taekwondo instructor?" She was jarred by his question and finally said, "No. I'm having an affair with Taekwondo." That captures the seduction of Taekwondo. I don't like Taekwondo—I love Taekwondo. Taekwondo has been internalized into my very being. It has become who I am, not just what I do. It's not just my sport—it's my calling.

More and more people began expressing their amazement at what I could do at my age. At this point, I was a blue belt and was beating black belts regularly. I trained three to four hours a day. I won the State of California competition in my division and qualified for the national letting these bad cells escape, grow, and cause disease and perhaps death. On the other hand, you can also have autoimmune dysfunction, where T-cells go haywire and kill off the good cells.

Monosaccharides

This new research in cellular communication has triggered the discovery of some new nutritional products I believe are the cutting-edge in cellular communication. This research has been ongoing for the last 20 years. It was brought to my attention a few years ago. We hear very little about this research that identified some essential elements of health. The research is finally starting to show up in major medical and university studies all over the United States and the world. That technology and research identified the importance of monosaccharides to human health. Monosaccharides are carbohydrates that are really single carbohydrates or single sugars. The research identified monosaccharides as the nutrients that actually go into the cell and facilitate cellular communication. The actual synthesis goes on in each cell is called glyco-synthesis or in other words monosaccharide synthesis. Monosaccharides or single sugars are extremely important to the whole function of the body. Within each cell and in inter cellular communication, monosaccharides are the medium for communication to take place. The exterior membrane of every cell in your body is composed of 80 percent monosaccharides. We know each cell interacts with other cells and they communicate with each other. In normal immune function the T-cell first hooks up with a cell. If it were a cancer cell it would be identified as such and the T-cell would organize the army to wipe it out. Unfortunately what is happening in our bodies is quite the opposite and that is why cancer is so rampant. Did you know not so many years ago one in five people died of cancer? That figure has gone way up. The most recent research reports half of the male population, 50 percent, that is one out of every two men, will die of cancer and one out of every three women will die of cancer. Today we know cancer is a huge and growing health concern. Perhaps one of the main reasons for the sudden rise in cancer is there aren't enough monosaccharides in our diets.

When T-cells hook up with cancer or other bad cells and there are no monosaccharides present, cellular communication does not take place or is inhibited. In many instances the T-cell will release that championship tournament. I didn't win that first national championship. I didn't even place, but it ignited a stronger desire to excel at that level. At Master Kim's encouragement I registered for an international tournament at the Olympic Training Center in Colorado Springs. I thought I would be able to fight my own age division and my own blue belt level. To this day I don't know if that was an honest mistake on his part, but I found out that if I wanted to compete, I had to compete with black belts in the 18-years-and-older division. I chose to compete. I actually won...what a high! I continued to compete, year after year. I sought more and more experience and knowledge. I went to Korea and trained a week with my Master's Master. If a training seminar was in my area with a visiting coach or high-level athlete, I participated. I began reading books and studying videotapes. My wardrobe now consisted of exercise clothes, Taekwondo uniforms and a few dresses for church and dates with my husband. Taekwondo became my self-expression in life. Ron encouraged and strongly supported me in my Taekwondo. After all, I was happy and thriving again. He was proud of me for my tenacity. He has always been athletic and fit. It was something we enjoyed together when we were dating. He loved that I got strong and fit again.

When Adam was old enough, I took him to Taekwondo class and enrolled him in tournaments. It has never been his self-expression like it is mine, but we had fun doing it together. I think having fun is his motivation more than any passion for Taekwondo. Many of the tournaments are in Las Vegas, Nevada, Los Angeles, California, and Orlando, Florida. That means fun. Adam will tell you he likes to compete and travel with the other students. Also he'll tell you the most important part of Taekwondo for him is having fun after the tournament. Snow skiing is much more his passion. I remember after recovering from his accident, cancer cell, because it can't identify it. An environment of insufficient monosaccharides becomes the genesis of dysfunction and disease. In autoimmune disease, the immune function goes haywire. A good cell hooks up with the T-cell and the T-cell starts killing it because of this dysfunction in communication. Once again the lack of monosaccharides my be the source of autoimmune disease. The amazing thing is when an ample supply of monosaccharides is introduced into the body through diet, research has concluded the body will readily accept them, regardless from where they came. When monosaccharides are present in the cells, normal cellular function and communication takes place. The T-cells are able to identify cancer cells; they rally the body and kill them off. Equally, when monosaccharides are introduced into the diet of a person with an autoimmune disease, lupus erythematosus for example, the T-cells will quit destroying the good cells. This should give you a basic understanding of the essential part healthy immune function plays in our wellness. It at least gives you a picture of how the immune system works and the importance of monosaccharides in that function.

If your immune system continues to malfunction or if you are not producing enough T-cells, eventually you will die. Everyone is now familiar with HIV or AIDS. Realizing a cure for this disease is very difficult because the HIV virus mutates so quickly they can't create a drug to fight it. HIV actually destroys T-cells or immune function. In that scenario, your immune function would continue to decline until you die of some other disease, because your body can no longer attack the destructive invaders. Immune function is measured by counting healthy T-cells. If the count is measured between four or five hundred, you have normal immune function. The immune function in patients with HIV may be as low as 10 or even 5 before they die. Is there an answer for HIV, cancer, lupus, and other advanced diseases?

Two doctors in Texas began researching the function and importance of monosaccharides. They started looking at immune function and they identified eight monosaccharides that could help the body and immune function. Over the last 150 years monosaccharides have all but disappeared from our diet. Monosaccharides used to be found in great abundance in fresh fruits and vegetables. There were many other sources and most were found in plant life. These researchers

Adam jumped off of high things—too high, like the top of the slide at the playground. It was as if he was trying to master falling. Skiing seems to be a continuation of that aspect. He skis double black-diamond level runs and is a daredevil on the slopes. He may take longer than anyone in history, though, to get his black belt. Adam started Taekwondo when he was four. He's now nineteen and is still a red belt. He's actually quite talented when he comes regularly. When he's going to a tournament, he'll cram with his practice and win the forms division. I also wish I had his natural timing and mental presence in sparring.

It seemed we just got through the roughest experience of our marriage when life delivered us another literal kick in the gut. Ron received a diagnosis of stomach cancer, a lymphoma, primary to the stomach. That is not a common tumor and is considered to be very aggressive. Without treatment, his life expectancy was no more than a year. The doctor wanted to first do surgery, then radiation treatment, and next chemotherapy. When he finished telling us all the statistics for Ron's chances of recovering from each of those therapies, Ron was at less than zero percent chance of survival. He decided not to surrender to the traditional medical course of treatment. If you have cancer, I'm not recommending you to do what he did—I'm just reporting on what Ron did. At that point, he didn't know what to do; he just knew what he wasn't going to do. I don't know what was more upsetting for me, imagining what he would go through with chemotherapy, radiation, and surgery or imagining him dying by not doing chemotherapy, radiation, and surgery. It wasn't my choice, though. It was wholly Ron's choice. In this car he was the driver and I was a passenger with no idea where we were going.

Ron began his healing journey by meditating for hours at a time. We talked about it. He told me he wanted to go on looked at many different plants, identifying and actually isolating the best sources for the eight monosaccharides. They started introducing them into people's diets. Miracles started happening. Many companies started trying to isolate the monosaccharides as they occurred in nature and they identified several plants had the highest occurrence of these monosaccharides. They then started figuring out processes to dehydrate these monosaccharides and ways to keep them alive through different techniques. They worked to develop a form that could be easily added to our diets. Monosaccharides in their purest form are available to the general public, but few people know where to find them.

Where do we find them? How can we take them? Do they really work? The most important part of the beginning of youthing has to be wellness or health. It must occur before you can really feel younger, you need to feel healthy. Many people get sick often, catching colds and flu back to back in the "season." Among other leading-edge new supplements, adding monosaccharides to your diet will build the immune system and you will see all forms of illness fade away. Three products have been developed into an optimal health system which allows your body to heal itself. You can refer to our website to get more in-depth information on these products. The most important thing to understand is that these three products have totally changed our health and could change yours.

In my own research, I found a company that is producing a pure monosaccharide product that has the highest concentration of monosaccharides available on the planet. You can get more information and find out how to order this product and other super supplements we use on our website, www.secretsofthefountainofyouth.com.

The first monosaccharide product I want to mention has been scientifically developed and tested over a 20 year period. Scientists actually tagged the product with radioisotopes and used it with test subjects to see if the body would accept a supplemental source of monosaccharides. Sure enough, it showed up throughout the body and throughout all the cells. It dramatically strengthens everything that has to do with immune function. God designed your body to heal itself of almost anything and adding monosaccharides to your modern monosaccharide-depleted diet will realign the body with that inherent design. This product has the highest occurrence of monosaccharides in existence. The interesting a quest to heal himself. He said he always wanted to travel our entire country in an RV and that if he was going to die, he wanted to die in one piece having a great time with his family. Adam was four years old at the time. We decided to sell our house in California, buy an RV and live Ron's dream. My only attachment to California was Taekwondo. I was sad at the prospect of having to leave Master Kim, who empowered me so much, but under the circumstances that was just the way it was. I made up my mind I would continue my study and practice of Taekwondo even as we traveled. I didn't know how I would do that; I just knew I would. Ron quit his job. I had built up quite a good income in a network marketing company, so we had a way to sustain ourselves. It was the first time I did network marketing, so the timing was miraculous. We bought a spectacular 35-foot RV. Neither one of us had ever driven or maintained one. My background in mechanics and electronics became very important. We both learned to drive it and I fixed most of what went wrong. When we drove into RV parks, heads turned. Our rig was striking. It had all the comforts of home. We sold or stored our belongings that didn't go with us. We set out and made it up as we went. We took turns driving. The one who wasn't driving played with Adam, reading to him, or teaching him to read. Adam was reading at a fifth-grade level when he started school two years later. He continues to be an above-average reader to this day. It was a great adventure and a time of incredible bonding as a family. As we traveled, we went to places where Ron could engage in his healing. He first had two weeks of therapy from a Korean doctor who was skilled in Suki, the art of opening the body's energy flow. Dr. Dohyun Choi lives in Dallas, Texas and was recommended to us by a friend. We left Dallas after two weeks of treatment. Ron had no stomach pain and was visibly better. He really looked sick thing is that this product contains all eight monosaccharides and the sources are all natural derivative. The developers of the product went out and researched plants and found the highest naturally occurring source of these monosaccharides and developed the system for preserving their nutritional value through a special dehydration process.

The second of these products is a breakthrough in nutrition technology and another of my weapons used to fight off sickness and disease. The supplement is composed of antioxidants that occur in high concentrations in fruits and vegetables. These nutrients are also essential in the healing process and to developing extraordinary health. The product is rich in these phytonutrients formulated in combination with the monosaccharides. The combination formula accelerates the body's absorption and utilization rate of both nutrients. This formulation is part of a powerful three-part system that not only boosts immune function but also provides the body with everything it needs to generate ultimate health and wellness.

Finally, the last product in this trio of life-altering supplements is all about balance. This is again a proprietary product formulated with phytonutrients, monosaccharides, and dioscorea, an element basic to all hormone production in the body. In order to have optimal health you must have hormone balance as well. In order to begin youthing and to have youthful feelings again, the body needs an absolute balance of natural ingredients to produce optimum health. None of these products are "uppers" or stimulants of any kind; they simply supply the resources the body uses to maintain optimum health and performance. With this formulation, the body finds a way of balancing itself and that balance is called modulation. In other words, if your body isn't producing enough of something this product helps the body to produce more of what it needs. If the body is producing too much of something it helps it to produce less. This enables the body to thrive in a homeostatic or balanced state of vibrant health.

The balance of all three products is the most synergistic formula of monosaccharides in existence. It provides the body with everything it needs to not only heal itself, but to deliver ultimate health. Imagine having inexhaustible energy and extraordinary health all the days of your life. Imagine having a body that could continue to do your dreams all before. Before he received a diagnosis of cancer, a friend came to visit and confessed to us later that when they left our house she'd told her husband Ron looked like he had cancer.

We then traveled to the Hippocrates Health Institute in West Palm Beach, Florida. There was a great RV park one mile from the Institute. We stayed there about three months. Ron ate only raw, organically grown vegetables, sprouts, and wheat grass juice. He swam in their ozone-purified swimming pool, took their classes in health and nutrition, and meditated. I found a Taekwondo school one mile from the park. We were both thriving.

When we traveled on the road, I get a local phone book as soon as we checked into an RV park and looked up the number of an Olympic-style Taekwondo school. I let the instructor know I was traveling and would like to come and take a class. They always enthusiastically invited me to train with them. To practice my kicks, I fastened a body shield to a tree with a bungee cord and whaled away. I could practice my technique either outside or in the laundry room of the RV parks. Many parks had exercise rooms. Sometimes I went to the exercise room of a local gym.

The silver lining in the cloud of leaving my beloved Taekwondo school and master instructor was that I had the opportunity to learn with hundreds of brilliant instructors all over the country for the next two years. What a rare opportunity. I still keep in touch with some of them. I learned to self-generate with my practice. I searched the Internet for competition opportunities on our path. I won state championships in about seven states. I continued to qualify, compete, and win in the national competitions for those two years.

After Ron completed his healing at the Hippocrates Health Institute, we hit the road again. It was winter and the days of your life. Making this possibility a reality is one of the main goals of this book.

That is just a brief mention of these products and what they do. Go to our website www.secretsofthefountainofyouth.com for more information. I'd also like to share with you my personal story of how I was introduced to this breakthrough nutritional technology. One of my neighbors was a distributor of these products and knew I was a runner. He knew I would try just about anything to keep my health and body going so I could run everyday — anything legal that is. He brought some samples over and said they were really good and I should try them. I thanked him, but quickly put them on a shelf and didn't think about them again for about three or four months. I had injured my right hip pretty badly and it started getting really painful. It seemed arthritic in nature. I guess the pounding from the running is what caused the pain. I couldn't run at all. I stayed off it for about six weeks but after considerable rest, I still couldn't run. I even started biking everyday so I wouldn't have the pounding from running. Although I did that for several weeks, it didn't help either. The debilitating hip pain continued. Every single time I tried to run, my hip really hurt. I remember once running about 200 meters and having to stop and walk home. I love to run. I was getting really depressed.

Then one day, I remembered the products my neighbor had given me. I went over to my neighbor's house to talk to him about them. He told me a friend of his had gotten the products for him. He also told me they would get me the whole line of products if I would really try them. I agreed and about a week later they came in the mail. They came in on Saturday and I remember taking them for the first time on a Sunday night. When I asked them how much to take, they told me these products were vegetable material and I could take as much as I want. They also told me people had administered these products to little, itsy bitsy, failure-to-thrive babies with profound results. Their point was the products were safe for anyone to use in any amount. I remember trying about a teaspoon of each. The third formulation of which I spoke was for adults and was in pill form, so I took a couple of those. I put them in my body, and the next day I actually felt better. My pain seemed to be gone, but I didn't want to take a chance so I rode my bike that day and I felt fine. I had absolutely no pain at all.

we wanted to stay in Florida. We took Adam to every tourist attraction in Florida, sometimes three times. We visited many great RV facilities. We met some great people and even expanded my networking business. Ron began to fly out and occasionally do some consulting work. Adam and I either hunkered down in a park and waited for him or drove to the next city where Ron rejoined us. As the weather improved in other areas of the country we set out for parts unknown. We stopped in my hometown of Atlanta, Georgia for another few months. Adam even went to kindergarten for a while. We left Atlanta and traveled back westward. We had some great experiences. We weren't in a hurry to get anywhere. When we saw a sign for something that looked interesting, we stopped. Sometimes, when we found a park that was particularly enjoyable and entertaining, we stayed a few weeks. We continued on our journey all over the country for almost two years.

Ron didn't die. He flew back to California for a follow-up exam and there was no remaining sign of cancer in his body. If he hadn't had his diagnosis confirmed at Stanford, they could have claimed it was a false positive diagnosis. They just say he had a spontaneous remission. They won't call it a cure, even though it's been 13 years now. His doctor didn't even want to know what he had done. He said he could never recommend it to any of his patients so he didn't need to know. He didn't even congratulate Ron. It seemed he was disappointed not to be right that Ron would die without the traditional medical treatments. I know some people are cured by chemotherapy but I can't help thinking that poisoning people just short of the point of death is not a good idea. Ron has been a real trailblazer for me. I know that if I ever got cancer I would do what he did. In his meditations he actually thanked the cancer for helping him get more in touch with his body. He asked

This is actually what happened to me. Getting well and helping your body to heal itself also depends on your body's individual inherent rate of response when it is provided what it needs. The depth and level of your physical fitness raises that level of response and determines how well and how fast wellness products will work. Here is the rest of my true story. I was training for a major marathon. That is a 26.2 mile race. It takes a lot of training. Most marathoners run at least 10 miles a day, every day for several months to prepare. After trying the biking thing on Monday and feeling great, on Tuesday I decided to take a short run and just go out and see if the products had really worked. If I still had pain, I would turn around and go home. I put on my running shoes and headed out the door. At that time I lived in the mountains. I began running up the street. I passed the good ol' 200 meter mark and felt fine. Everywhere I ran was hilly so I climbed a few hills and amazingly felt no pain. It seemed like every time I thought I might start to have pain, I would feel a tingling sensation and the sense of impending pain would go away. That's the best way I can think of to describe it. I kept running and running. I had been biking for a few weeks, so the monosaccharide products were not necessarily the thing that made me run so fast; it's just they made my pain go away. I ran and ran and ran some more. I got about four miles into my run and I was thinking, "Gosh, this is amazing." I decided to push it a few more miles. I decided to try some one-mile repeats in order to really test the stuff out. I remember running a one-mile repeat, which is going as fast as you can go for one interval and alternating it with a slower-paced mile. I ran about a five-minute plus mile. Then I jogged a mile and I still felt okay. I tried another one and had the same result, no pain and it was a fast mile. I tried another five-minute-mile, then again slowed the pace to rest the next mile. I again ran another mile repeat. I ran three mile repeats all less than six minutes per mile. I couldn't believe it. Anyway, I got back home and discovered I had run eleven miles. I have never had any more hip pain from that day on.

I don't think that was coincidental and my training went extremely well from that point on. I was able to compete in my marathon, which took place a couple of months later. In the meantime, the neighbor who had given me the product told me to write down an account of every ache, pain, or any physical issue I'd had before I started taking the products. I wrote them all down and I remember specifically what the main things were.

it what it was trying to teach him. He asked his body what it needed to heal itself. He listened and followed his intuitions. Almost miraculously we would come across the next step by accident or we met a person who heard of something that would strengthen his immune system or have a healing effect. We learned that what you eat makes a huge difference in your vitality and health. Ron had also been working so much he was literally missing the experience of marriage and family. He loved his work and never would have quit. Maybe we create honorable ways to quit by getting sick. Sometimes tragedies alter the course of our lives like nothing else can. Adam's accident gave me Taekwondo and through confronting his own death, Ron learned to really live

It's been 18 years since I started taking classes in Taekwondo. I am now a WTF-certified, fifth-degree master instructor. I am a 15-time U.S. national champion and have won gold in five international competitions. In 2004, I was inducted into the U.S. Martial Arts Hall of Fame as International Female Competitor of the year. At the 2004 Korea Open I was awarded the Hwarang Warrior Female Competitor of the Year. I taught and coached over 100 students to national and international titles and produced the first Taekwondo national team member from Utah. Three of my students now own their own Taekwondo studios. I continue to teach and train two to three hours a day. In 2006, I broke the weightlifting world record in the chest press for my age and weight division. I took up weightlifting along with my husband, who led the way in power lifting and holds the national record in his division. In 2006, he also won his division in the world competition and I know he won't stop until he gets the world record.

Ron and I are engaged in the quest of an anti-aging breakthrough for others as well as for ourselves. We learned

My left eye had been twitching for about a year or two. It would twitch for about a week, usually three or four days out of the week. It would bug the heck out of me and I just thought there was nothing I could do about it, so I just tried getting more rest. What I did notice was after I started taking the products it went away. I can't say exactly when, but within at least a few weeks or a month or so it just went away. I noticed it was gone and it has never come back. About a year later, when I had been really tired after a particularly rough week, I looked in the mirror and saw my eye twitch once. I ran to the cupboard and took a whole bunch of monosaccharides. The twitching stopped and has never come back.

One of the most miraculous things I discovered after taking the products had to do with a chronic pain I had in my right shoulder, just above the shoulder blade. At times it was really severe. It was probably remnant damage from a bicycle wreck that had happened six or seven years earlier. It bothered me pretty much about half the time. If I slept wrong in any way, it came on and I suffered with it sometimes two, three, or four weeks. I tried to work it out and strengthen it. I went to the chiropractor or tried some new therapy. Sometimes it went on for months. One time I got into my car and was not able to turn my head around to look as I backed out of the driveway. It was nighttime and I checked both of my mirrors and it looked clear. I put the car in reverse, stepped on the gas, and ka-boom — I had hit my son-in-law's car and smashed in the whole rear end. He had backed into the driveway and our cars were back to back. I couldn't see his car in my mirrors and I just totally smacked the back end of his car. That is how bad my shoulder pain was. There honestly were times when it was so debilitating I had to hold my arm up the whole day. I would press hard on the painful spot over and over again, trying to relieve the pain, just to get through the day.

As I took the monosaccharide products, I noticed my chronic pain just went away and never came back. What is miraculous about the products is you don't notice what it does until after whatever was bothering you disappears. I remember it took about a month before I noticed I had no pain. It's been seven or eight years now. No matter how I sleep, it has never returned. Many other little aches, pains and nagging discomforts have gone away. I can't even begin to tell you how great I feel.

so much on this journey. Ron talks about getting cancer as being a "blessing" in his life. We continue to study and expand our knowledge in human nutrition for health and now optimum performance. Ron does a juice fast one day a week. The juices are, of course, freshly extracted from organically grown fruits or vegetables. He does a lot more juicing than I do. Both of us eat a lot of fresh raw vegetable salads. I eat more chicken and turkey than Ron does. I eat basically vegan and yet am not religious about it. I sometimes eat animal meat or fish for convenience or to be polite when someone has prepared it for me. Ron gets his protein from soy products and sprouts. I pay attention to food combinations more than he does. I eat proteins, fats, and fibrous carbohydrates, i.e., color vegetables together. I eat whole grains but only with fibrous carbs, not ever with meat or fats. I eat fruits alone and only early in the day. As we age, the enzymes our body produces are greatly diminished. This is probably why we tend to get more and more digestive distress as we get older. Raw vegetables naturally have enzymes that support digestion and something raw should be eaten with every meal. You can also take enzyme supplements. I have eliminated high glycemic index foods such as corn, carrots, beets, and all refined starchy carbohydrates. Although I occasionally indulge in "death by chocolate," it is now a widely known fact that sugar is more of a culprit in fattening up Americans than a diet high in fat. I think it's important to regulate a minimal animal fat intake if you want to manage a healthy body weight and body fat ratio. I also maintain an appropriate intake of good fats, such as foods and oils rich in essential fatty acids. It is very important to your health and weight management to drink at least three quarts of pure water every day. I know how hard that is for me to do even with my commitment to do it. That

Talk about youthing or about a miracle — the ability to function flawlessly every single day is what youthing is all about. People are amazed at my function. I am 57 years old and I'm able to go out anytime, even without serious training, and run a sub-six-minute mile. My level of overall health is extraordinary for my age. Another important result I have noticed since supplementing with monosaccharides is the quality of my sleep. I used to wake up several times at night; since being on the three products, I have slept soundly every single night and I don't need as much sleep as I used to. I think it has to do with toxicity. Sleep experts say sleep is simply a way to remove toxicity from the body and people who get more sleep have less risk of cancer. There are many other benefits regarding sufficient sleep that are extremely important. I don't even know the scientific evidence about this product as it relates to sleep; I only know the quality of sleep you get when you are on these products can totally change your life and create an unbelievably healthy body.

I'm going to give you just a couple of other examples of the effectiveness of monosaccharides. First of all, my wife has also had remarkable experiences with the products. When she was about thirty years old, she went through some sort of bodily change. It's pretty interesting. We were traveling back East and she started sneezing and sneezing. She thought she had a cold. She had a scratchy throat and itchy eyes. It kept going on and on and on. Finally, she went to the doctor and he diagnosed her with hay fever. She wondered how in the world this could just happen at her age. She had never had hay fever or any other respiratory problem in all her life. But sure enough, every time the pollen count of certain plants went up, she had really bad hay fever. It was miserable; her eyes teared up, her nose ran, and she would wheezed. It was really severe. I even thought maybe in her old age she would become asthmatic like her sister. Her sister had been asthmatic ever since she was a little child; but my wife had just developed this allergy when she was thirty-something years old. That was the odd part to me. It continued for years and years. I remember spending hundreds and hundreds of dollars on medications. They would come out with something new and my wife would try it. Finally, once I had experienced the success with my running on the monosaccharide products, we asked my neighbor about hay fever and other illnesses. He told us it should improve anything the body does leads me to suspect most people are actually dehydrated. Drinking sodas, even if they are sugar free, also complicates dehydration. What's worse is that sodas actually cause destruction of red blood cells almost immediately after you drink one. I personally saw this demonstrated with a dark field microscopy preparation of my own blood. I never drank another soda after that demonstration. Until that experience, I always rationalized that "sodas were my only vice and they couldn't be that bad." They are that bad. The body will hang onto fat if it is under-hydrated. As the body lets go of fat with adequate hydration, it also lets go of toxins stored in fat cells. "Water is life" is the long and short of my message.

As we age, our bodies process food and utilize nutrients with less efficiency. I am a firm believer that my ability to physically perform at the level of a 30-year-old is also a result of extensive supplementation. I take the highest-quality, leading-edge-technology vitamin and mineral formulations. Every day I take formulations that support the bone and cartilage, stimulates and fortifies the immune-system, provides cardiovascular support, and enhances brain functions. Recently I was introduced to the most important supplement in my commitment to have a body that can keep up with my goals and dreams. It is a literal breakthrough in the science and technology of anti-aging. The Synergy Worldwide company markets this product. They are based in Orem, Utah, 15 minutes from my home. A famous doctor, who is best known for her research in glycemic index, developed this product. She worked 20 years with a famous doctor, who is best known for his breakthrough research with vitamin C. This anti-aging innovation is called ProArgi-9 and is proven in hundreds of thousands of case studies to stimulate your own pituitary gland to produce the human growth hormone to heal itself. My wife decided to try out the monosaccharide products and what she experienced was miraculous. From day one her hay fever went away. We discovered another product that combines the monosaccharides with Vitamin C (natural source). We will call this Product D and she took this in conjunction with Products A, B, and C and her hay fever went away and never returned. It was not just relieved — it went away. Again we don't remember exactly how long it took; we just know it went away. We have been on these leading-edge products consistently for the past several years. Since that time we have seen countless miraculous changes, and none of the problems have returned.

There have been periods in which we stopped taking the products for about a month to six weeks, but none of the symptoms returned. One of the other things that happened to my wife on the products is her eyesight improved. As strange as this may sound, some people have mentioned that the products improved their eyesight. Ever since my wife was a little girl in elementary school — I think since the first grade — she had worn glasses with very thick lenses. As she took more and more of the products, her prescription actually changed. Over time, her eyesight got better and better. Somehow I didn't mind spending the money on new glasses. I don't know what youthing is to you but when you can take something that gets rid of your hay fever and improves your eyesight, it's so unheard of it's miraculous. If that isn't youthing I don't know what is. Without her glasses, she couldn't even read the clock on the night stand next to the bed. It was a digital clock with numbers about an inch and a half high. She can see that clock clearly. Actually in order for her to focus close up, even though she has bifocals to read the small print on anything she has to take off her glasses. In order to read the little tiny print nobody can read, not even young people, she has to read without her glasses. With her glasses on she can't read that small print. I'm still amazed she takes her glasses off and reads it with her natural vision. Those are just a few benefits and we could go on and on, but let's just use those as an example of nutritionals help with the many, many health issues related to aging and youthing. Definitely go to our website www.secretsofthefountainofyouth.com and find out how to get and use these amazing monosaccharides. Youthing begins with health and well-being.

levels of a 16-year-old. At the ripe old age of 23 our bodies are programmed to begin slowing their production of the human growth hormone. Every decade the body produces less human growth hormone. The absence of human growth hormone is responsible for almost all the effects of aging. Toxins in our environment and even the toxic effects of stress are also big culprits in aging the human body. When my husband and I moved to a home in the mountains in Sundance, Utah, we eliminated many of the biggest sources of environmental stresses on the body. We have clean air, and the kind of spring water that is bottled and sold commercially comes out of our tap. The natural beauty of our environment is also nurturing for body and soul.

You can check our website www.secretsofthefountainofyouth.com to get all the names of my supplement regimen and how you can purchase them. Synergy also has a breakthrough anti-aging skin care line that works to reverse the effects of aging on the skin. I always use leading-edge skin care products, but I never used one that actually caused the deep lines around my mouth and eyes to fill in and look 200 percent smoother. Again it is a signaling technology. Enzymes are carried to the dermis layer of the skin by bi-pyramidal liposomes and stimulate your fibrocytes to again produce the kind and levels of collagen they produced when you were young. The result is that lines and wrinkles begin to fill in permanently over the months as you use it. People regularly guess I'm in my late 40s and I am 63. Performance is of the utmost importance to me, but looking good is a close second.

The most important benefit of sustained extensive supplementation (although looking good is probably my favorite benefit) has been that my energy, strength, speed, flexibility, mental focus, recovery, and agility have more than doubled. My bone density is 125% of normal for a

There are so many things we could say about disease and monosaccharides. We could actually quote people who have walked away from every known disease. People on monosaccharides have walked away from Lou Gehrig's disease, most types of cancer, Crohn's disease, multiple sclerosis, and muscular dystrophy. There's a medical doctor up in Canada who had an aneurysm and that resulted in some permanent brain damage. He was in the dining room and felt this thing strike him, felt like a sledgehammer hitting his head. As he went down, he screamed to his wife that it was an aneurysm and they rushed him to the hospital. They managed to save his life but he was an invalid for nine years. One day his maid came to him and told him to try the monosaccharide products. He did and within two years he was a normally functioning human being. I personally heard him speak. I actually saw him in person. He said, "I know I'm not supposed to say this, but as a doctor I'm just going to say what I think this is in just two words *genetic repair*." He said there is no other explanation for his being able to overcome the damage done years before to his brain, to fully recover, and be a normally functioning human being.

We could go on and on and on. This is amazing stuff and it really does reverse the "normal" decline of aging and propel your body into youthing. Bodies thrive when they are not fighting illness. Bodies and human beings thrive when they don't have any aches and pains. I have no aches and pains. When I hit fifty I started getting pains everywhere. I started to literally fall apart. My knee was hurting and, as I mentioned before, my hip was totally whacked; I had aches and pains all over. Much of this was probably arthritic in nature. Like me, many other people have experienced health miracles. They just walked away from debilitating conditions such as hands malformed by rheumatoid arthritis and almost anything else you can imagine. Still, it's my experience I can share with you and I have become a one hundred percent normally functioning human being. No, at 57 years old it's really abnormal to function and perform as I do. I'm out to have vital health and aliveness be normal for everyone.

I'll share a story with you about a guy who was dying of AIDS. His immune count was down so low they couldn't even measure it. I believe it was single digit. He was very close to dying. One of his friends sneaked him the products because the doctors would not allow him to take nutritionals

25 year-old woman and my antioxidant level is 47,500. I am enjoying a vitality that is uncommon today for a 25-year-old. I have not had a full-blown cold, flu, or any other illness in over 10 years. If I start to get a cold or flu, my regimen is to take a homeopathic remedy called Alpha CF (Cold-Flu) or if it is strictly flu symptoms with gastric distress and body aches, I take another homeopathic remedy called Oscillococcinum. I add the herbal remedies of echinacea, golden seal, cayenne, and garlic. Every hour for the first eight hours and three times a day thereafter for three days, I add 1,000 mg of vitamin C as a buffered, pleasant-tasting drink called Emergen-C. If you do this regimen at the first sign of symptoms they disappear within a few hours. As a world-class competitive athlete, I don't worry about not feeling my best on competition day. In 2000, when I didn't make the U.S. Taekwondo Olympic team, I gave up my Olympic dream and retired from competition in the 18 years-and-over division in sparring. I have maintained my national first place title in the technique division for the last 15 years.

I love my husband, I love my life, and I love having a body that can do my dream of participating in world-class athletic competitions. I also think it's my purpose to create a new reality around aging. Living from a mission and a vision is a source of aliveness. It keeps me going and it keeps me "youthing" instead of aging. It's easy to take it easy, but that is the path that will give me the commonly accepted fate of the decline of aging. I love being an inspiration and a model that the decline of aging is a choice and that anyone can be alive vitally his entire life. Supplementation made a huge contribution to me, and I have an intention to contribute to others by teaching them the path I've discovered to extraordinary vitality.

and within five to six months his immune count was up over 500. Even though he carried the virus, his HIV was now non-detectable! They were going to do a huge study on HIV and the use of monosaccharides at one of the universities in Dallas and it got stopped for some reason. Remember, monosaccharides come from plants such as vegetables. If you know someone who is ill it sure won't hurt him or her to try them. The proof that monosaccharides are a phenomenal supplement and they really work is to put them in your body and see what they do.

Another important application is for kids with ADHD and that sort of stuff. Even kids with HIV can try this. The company makes these little gummy bears that contain monosaccharides and gives them to kids all over the world. They actually have a Christian ministry that makes these supplements available to children throughout the world. They went to Romania and to one of the orphanages referred to as an HIV orphanage. You may know there was an HIV epidemic in Romania. Many HIV-positive children are in these orphanages because their parents have died of HIV. Some time ago I heard a woman speak at a conference and she said they had, I can't remember exactly, but about a hundred and something kids in their orphanage and during the winter half of the children would have to be hospitalized with different illnesses and many of them would die. I don't remember the exact statistics but it was tragically significant. She said one day these wonderful foreigners came to her and told her to start feeding these magic gummy bears to the kids. They told her to give the older children lots of bears and to give the younger ones less. The next winter, I think it was in 1999, she said zero kids went to the hospital and none died.

Another little project up in Canada took place with some teachers. These two teachers took on all the ADHD kids in their elementary school. If you could imagine having 15 or more ADHD kids in one room — it seems unimaginable. These teachers heard about the gummy bears, and contacted the Christian ministry, and received the gummy bears. They started giving them to these kids. They said within 30 days they had a normally functioning classroom without any kids bouncing off the walls or misbehaving. Once again, if you don't believe it, just try it. They are gummy bears with monosaccharides. We can put you in contact with

My husband and I have a great adventure every year. One of my heroes and sources of inspiration is the ancient Japanese samurai warrior, Miyamoto Musashi. I've read his *Book of Five Rings* over and over. His life story was a book of about 700 pages and I literally lost myself in it for weeks. It was one of the best adventures of my life to travel to his birthplace in Japan. Ron and I had no idea how to get there from Tokyo and we didn't speak Japanese. In the train station in Kyoto we found a history major volunteering at the information booth. He just happened to speak perfect English. He gave us instructions and we set off. We started on the "Bullet Train" of Japan and transferred to smaller and smaller trains until we were on a one-car train full of students. We got off at Musashi Station. It was out in the middle of nowhere. When I said "Miyamoto Musashi" as a question, a local drunk nodded and escorted us to the birthplace and dojo (training school for martial arts), jabbering in Japanese as if we knew what he was talking about.

After walking about three blocks we found the dojo, the museum, and the home in which Musashi was born. When we arrived I looked into the dojo to find a sword-testing going on. It was magical. I enjoyed it. Then we found out they had only one room left at the hostel. It was inexpensive and included our meals. They had a great Japanese bath there and reed floors. We slept in the traditional Japanese style on mats on the floor. The next day we hiked some of the most beautiful hillside trails I imagined Musashi had roamed centuries before. What an adventure. Contrast that to going on a choreographed tour. Plan adventures into your life. Adventures cause growth. Adventures are the environment of aliveness.

We trekked the Inca Trail in Peru last summer. It was the most grueling ordeal of my life. After hiking straight literally hundreds of people who have tried this. Once again, go to our website www.secretsofthefountainofyouth.com for all the information you need.

Even with our highly evolved modern science, we don't know all the physical ramifications of the food we eat or all that's caused by not eating certain foods. It can only be supposed the high occurrence of type II diabetes, ADHD, and all the other health issues I've mentioned could be the result of our diet. One of the reasons could be because our bodies are lacking the eight essential monosaccharides. These single sugars are key elements in our health and in optimally sustaining all of our body functions. I for one have used them and have experienced their miraculous effects on my own health and well-being. I can't say enough about them. In your journey of youthing, I suggest you incorporate these products into your diet and see what they can do for you. I believe you could be in any state of health, maybe even on your deathbed with whatever illness, and I'll just say it again — you can't go wrong trying monosaccharides. They are vegetable material, for goodness sake. They can't do you any harm, and you may have some miracles of your own to share. There is no absolute guarantee monosaccharides will reverse everything you have or that they will make you 100 percent better. I just know they have been one of the most important supplements I've ever personally experienced for health and well-being.

I remember meeting a lady who handed me one of her business cards. On one side she had listed all the health problems she had previously had. She said she had been so ill she was ready to die. She had so many things wrong with her that this big long list of ailments took up one side of this card and it was in really fine print. On the back of the card she had a list of every medication she had been on. She'd started taking the monosaccharides and experienced some immediate effects. At that time she thought it would be impossible for her to totally recover. She told her entire extended family she might not live because she was so far gone but she was going to try these monosaccharides. Well, needless to say, she lived. She was passing out this business card a year later and she said that she was free of every single health problem on the front of the card and she was not taking a single medication listed on the back. If up for six hours to a 16,000 foot summit where we froze our buns off, our reward was to hike down for another four hours...in the rain and cold! Don't believe travel agents who tell you there's a dry season in Peru! At night when I had to use "the facilities," I knew when I went outside I would sink up to the top of my ankles in mud. I argued with myself long and hard before necessity won out. We hiked like that for four days in a row. Nice vacation! It was a great adventure though, and now that I've done it, I feel very proud of myself. Coming around that last mountain and looking down on Machu Picchu was worth it all. The hot shower at an incredibly beautiful hotel at the end of the trail was pure ecstasy. The only other person even close to my age (50) didn't make it two days and had to take the train. My old mountain goat of a husband did it easily. He hiked on ahead and came back and did it again with me. He is amazing!

I most desire to be a model and inspiration for people of any age to take the risk of passionate engagement in impossible dreams and do the daily work to realize their dreams and goals at any age. Create and engage in adventures. We are most alive when we are living an adventure. We need to move our bodies strenuously every day. Dream the impossible dream and then live it like it was possible. Seek out and use breakthrough technology in the science of anti-aging. Your project is to have a body that can do your dreams at any age. Do the work to complete your self-limiting patterns and create a life you will love. I did. It's just a matter of engaging in the opportunities that present themselves to you. Take action every day to realize your dream. What else are you doing for the rest of your life? Use my path or invent one of your own. Live from a commitment to live a life fully expressed and the way will appear. Being your best self is not the "right" way to live.

anyone complains about an ailment to her, she recommends they take monosaccharides, even to prevent these health problems before they start. So, there you have it. That is what these monosaccharides can do.

I went to a couple of conferences about monosaccharides because I wanted to find out more about them. I was amazed at what I saw. In Seattle I saw people parade across the stage many of them accompanied by their doctor's and talk about walking away from pretty much every disease you could possibly think of. They told what happened to them and how it went away once they got on monosaccharides. I really wish every human being could see what I saw. That is part of the reason why I am writing this book. Monosaccharides are safe, scientifically researched products with amazing health benefits. Monosaccharides are the gateway to health and the beginning of youthing. You cannot grow younger without your health, without being free of disease or sickness.

Monosaccharides can help you on the path to youthing, on the road to feeling better, on the road to having energy and all the things you need to have the vigor and vitality of your youth return to you. Even if you are 100 percent healthy and don't have any aches or pains, monosaccharides can help you maintain what you already have. Go to our website www.secretsofthefountainofyouth.com for more information on monosaccharides.

Other Well Kept Secrets

This chapter wouldn't be complete without a few of the other secrets associated with health and well-being. The reason these are secrets is that most people don't know about them. This could be the first time you have ever read about some of these things. These are truly secrets that could make a profound difference in the quality of your life. Please spread the word once you know what we know.

In years past, wealth and health were defined by the colloquial saying called "being born with a silver spoon in your mouth." Today, silver spoon usually refers to a spoiled kid. During the days of the black plague in England, the babies of wealthy families were fed with a traditional baby spoon, which was made out of pure silver. Those babies didn't die of the black plague. Why didn't they die? It was because the little microscopic

It's one possible way to live. It just happens to be the most satisfying, the most alive, and the most fun. It is a source that waters the fountain of youth.

pieces of silver that broke off from the silver spoon with which they were fed had some unique germicidal qualities; the presence of the silver in their systems actually killed bacteria and viruses. That's the rest of the story about "being born with a silver spoon in your mouth."

Years of research in the early twentieth century revealed the absolute proof that microscopic silver is 100 percent safe for the body and that the small particles of silver (colloidal silver) were antibiotic and antiviral in nature. The research was hidden and not made public.

Colloidal Silver

Today, there is a product that uses this technology and it is called colloidal silver. About fifteen years ago, there was a renewed interest in this type of silver, which was developed by people who had learned about it from their ancestors. Their ancestors had produced colloidal silver on stove tops and used it as a remedy for bacterial and viral infections. Again, colloidal silver is microscopic and it doesn't harm you, nor can you overdose on it. In fact, it is safe for everyone, even babies.

Many years ago, I had tried a form of colloidal silver. When I got sick, with either a viral or bacterial infection, I sprayed the liquid silver in my mouth, breathed it in through my nose, or took it internally. The result was my illness disappeared almost immediately. My sickness just went away. It was really interesting. When you go to doctors for the flu, they always say, "It's a virus and there's really nothing we can do. Just go home, take some over-the-counter flu medications for the symptoms, and it will go away in a week or two. Next year make sure you get a flu shot." But, if it's bacterial in nature, like bronchitis or any other sort of bacteria, they give antibiotics and everybody knows the danger of antibiotics as far as the after-effects and what can happen to your system.

Actual studies have recently been done on colloidal silver. These studies were done at the University of Illinois, Institute of Technology, and Brigham Young University. I have actually read the reports and have seen these studies. In these reports, they compared the effectiveness of colloidal silver to the leading antibiotics. The conclusion of this research was colloidal silver was equal to or better than five of the leading brands of antibiotics and there were no known side effects.

NOTES

Once again, colloidal silver consists simply of little microscopic particles of silver suspended in water. It's just plain water with microscopic silver particles suspended in it. What they found in hospital tests was colloidal silver kills off strep and staph and all other known harmful bacteria and viruses when in contact for at least two to three minutes. They've even shown higher concentrates of colloidal silver will kill off anthrax and anthrax spores. It's a great water purifier as well. We spray it in our house regularly, especially when someone with a cold or something contagious has visited.

The colloidal silver needs to stay in contact with the germ two to three minutes in order to kill it. Because I am taking the monosaccharide products and use silver whenever I even feel like I am starting to get ill, I have not been sick or spent a day in bed for more than ten years. Could this be the cure for the common cold and flu? Try it and see for yourself.

I never get a full-blown cold, but if I do feel something is coming on, I start spraying the colloidal silver. Within a few hours the symptoms are gone. I haven't kept an exact journal, but that's the way it works. Colloidal silver can be sprayed on surfaces to kill off germs. It kills any kind of bacteria or virus and doesn't harm any of the good bacteria, unlike antibiotics. We always have it in our cupboard. I have a bottle of the highly concentrated formulation if I think I need some extra strength or in case there is ever danger of anthrax or any other kind of germ warfare. The website www.secretsofthefountainofyouth.com has more detailed information on this amazing product and the way to get it.

pH Balance and Ionized Coral Calcium

There is another function of the human body that has a direct influence on the immune system. It is called pH balance. The pH, or acid-base balance of the human body has a profound influence on overall health. Did you know cancer and most diseases thrive in an acidic environment and actually cannot exist in an alkaline environment? Take a guess about whether our bodies are naturally alkaline or acidic. What are we, naturally acidic or naturally alkaline? You guessed it — we are alkaline. Guess what? The body can pretty much fight off disease and keep us healthy if we maintain the pH balance and keep our bodies alkaline.

NOTES

Some food and drinks are highly acidic and, if possible, should be completely eliminated from your diet.

All alcohol has a pH of 4.0, coffee, 4.0, carbonated drinks, 4.0, and orange juice, 4.0.

We know this information will disappoint many people but the cold hard facts do sometimes. The very best thing for you to drink is pure water. It takes your body a day to recover normal pH after taking in these acid-producing products. I suggest you give your body a break and stay away from them.

Some foods and drinks are alkaline and will contribute to proper pH balance. They should be included in your diet.

Eat vegetables, especially fresh organic vegetables, fruits, especially fresh organic fruits, and drink plenty of pure water

Are we suggesting you need to be a vegetarian? No, we're not suggesting that at all. Just eat lots of fresh fruits and vegetables if you want to be healthy. Meats and dairy products are fine in moderation if you choose.

If you were observant and looked at the heading of this section, you probably noticed it said something about ionized coral calcium. We are going to offer a brief explanation about a secret of the fountain of youth. In Okinawa, Japan, people live a whole lot longer on the average than anywhere else in the world. There, life expectancy exceeds that of any place on the planet by about ten years. Some scientists went there to try and find out why. They did. They spent several years there looking for a scientific explanation for the Okinawan longevity. What they found was very interesting.

Okinawan water was not only alkaline but provided the residents with some pretty amazing nutrients. The water was being taken out of the ground and the coral-based aquifers were full of ionized calcium and 73 other ionized minerals. What does *ionized* mean? That is the first thing I asked myself. So, here goes another unsophisticated explanation for the nonscientist like me. Minerals come in all sizes. Most of them are actually rather large in microscopic terms. Let me first tell you that for millions of years these coral reefs have been broken down by wave action and the alkaline sea environment. The calcium and the other 73 minerals found in Okinawan water were actually ionized over millions of years. What that means, for example, is the calcium found in a normal

CHAPTER 2

THE PRINCIPLES OF MASTERY

Be the Master of Your Own Life

Living "the mastery game" is one possible way to live life. It's not the right way. It's not even the best way. It's a choice. Do I climb the mountain? Do I lie on the beach? The mastery game always has risk and adventure as a fundamental element. I consider those engaged in the mastery game to be modern-day warriors...even if they are dancers or doctors. What is the mastery game? I think the best thing I can do is describe the elements I have identified through experience. I have been on this path for almost 20 years, and being a certified master instructor, and a coach in the Olympic sport of Taekwondo, I'm actually addressed as "Master Zeller." The purpose of mastering the martial arts is not only to develop a high level of skill in the techniques, calcium tablet you buy from the store is huge. If we were to say the calcium molecule in those tablets was the size of a softball, for example, a colloidal calcium molecule would be the size of a grain of sand and the ionized calcium would be ten thousand times smaller than the colloidal calcium and exponentially smaller than the calcium tablet molecule. In other words, all the calcium and the 73 other ionized minerals are able to get into your system at a much higher and faster rate than any other form of mineral in existence because the particle size is microscopic.

The other key factor about coral calcium is that because of the smallness of the particles, it is a key factor in changing everything it comes in contact with, to alkaline. It has been scientifically proven to help change your body to an alkaline condition again. It is an amazing electrolyte as well. On Okinawa cancer and heart disease are almost unknown. The well-known research in Okinawa has spread to mainland Japan where over a quarter of the population now consumes coral calcium.

As a runner I started using this product and found after extensive running workouts, I recovered faster and did not fatigue as quickly. It is an amazing electrolyte. It is available in tablet form or in tea bags with actual coral sand in them that you put in your drinking water. Remember there are 73 other trace minerals with the calcium, creating a remedy that brings health and well-being, not to mention living a healthy life to a ripe old age like the Okinawan people. This is definitely a secret to the fountain of youth. For more detailed information you know where to go. www.secretsofthefountainofyouth.com.

I honestly believe we have presented some of the products that can eliminate all illness out of your life and restore your health. That is certainly the key to and the beginning point of all talks of youthing or being youthful and one of the several keys you have just received to the fountain of youth.

There you have it. These are some of the most cutting-edge ideas and products in existence to get you well and going strong. Remember, it is impossible for you to even start a youthing journey without at least a basic level of wellness. The information we have provided you with in this chapter is what can help you get there. The website will expand on all these things and post any updates as we find new or improved products and protocols to help you get your health back or to get to a state that allows you to be on the path to youthing.

but it is to master oneself and life itself. Many achieve the high level of skill, yet I have met only a few who embody the true spirit of Master. What does it take to be a Master and not just achieve the rank? Why engage in mastery? Again I look to my own experience. Instead of the normal decline of aging associated with middle age, I began to experience a sense of power, a physical prowess usually only associated with youth, being vitally alive and satisfied with who I am like never before in my life. I found myself making a significant contribution to the lives of others, directing them to discovering their own significance and magnificence. It's a very enriching way to live life.

Mastery is not just accomplishment, although accomplishment does coexist with mastery. The mastery game is seductive once you discover its treasures. Being bored to death is a terrible death. Every year engaged in mastery is to have a new life. I suspect if we engaged in mastery in marriage there would be no divorce. It would always be growing and new. In mastery we discover a self that would never have been without the mastery game. If you can use my story to generate the best possible life for yourself, it would be one of my greatest satisfactions in life-and in yours. As I share my story and lessons with you, remember this book is not about me, it's about you and a new possibility for your life. To play the mastery game is to transform the possibility of who you can be…at any age! One thing I know from my experiences over the last 20 years is that we really have no idea of what we can do. Even with some of our greatest achievements, we still settle for what we know we can do. Greatness lies in doing what we don't know we can do. Many people are good and have good lives and comfortably settle into cruise control. There is certainly nothing wrong with living a good life. Sometimes, though, greatness is sacrificed for good. Greatness exists in

CHAPTER 5

WEIGHT LOSS FAT AND MUSCLE BALANCE

The New You

One of the most recent developments and the biggest news about health and nutrition in this country and all over the world is the state of obesity in the U.S. The latest statistics show about a third of the country is overweight. Overweight means you weigh more than 15 pounds over what your normal weight should be for your height and age. Another third of the country is obese. Obesity means weighing thirty pounds or more over what you should be. What an alarming state of being… the United States is two-thirds overweight or obese. That is scary! Much of this has to do with our lifestyles, our diets, the things we snack on, the things we do every day and the amount of stress we are consistently enduring.

One of the main contributing factors is our country is moving away from a once active lifestyle as seen in Europe and Japan. In these countries everyone still walks everywhere they go; we don't do anymore.

every person. Sure we all have some talent to one degree or another, but maybe greatness is common to us all if we were to take a certain path in life. What's the access?

Some things take time. Mastery is one of them. It's a path, a way, not a destination. It's like the skin horse in *The Velveteen Rabbit*. After a long, long time, he says, one day you just become real! How do you know you are playing the mastery game? You don't know when you enter the game, but there are certain elements that transform ordinary life into the mastery game. I think that in every life come small, sometimes only cubic centimeters, of chance, certain rare opportunities that when recognized and seized, change the possibility of our lives. The only difference between an ordinary person and a "warrior" is that the warrior, or the extraordinary person, not only recognizes this fleeting opportunity but has the audacity to seize it as his own and the tenacity to passionately engage in the opportunity over whatever time it takes to develop the opportunity to its fullest. It is in the passionate engagement that these particular opportunities over time transform us. It's not about the accomplishments or the winning - it's about what happens to you on the path. One of my favorite quotes is by Rainer Maria Rilke. He says:

If I don't manage to fly, someone else will.
The Spirit wants only that there be flying.
As for who happens to do it,
In that He has only a passing interest.

Many people, upon being offered an opportunity, will look to their comfort zone or look to see if they think they can reasonably do it, and will let their considerations of what they think they can comfortably do determine their destiny. When I accepted the opportunity to engage in the

Here, everybody has a car. In Japan many people take trains and buses, yet once they get off the bus or the train they often walk quite a distance to get to the office or their home. What about going out for lunch? Do we walk or drive? Unequivocally, one of the largest contributors to our fat and obesity epidemic, is the fact we don't exercise enough in our daily routines. Most of us sit behind desks all day, get in our car, and walk maybe a hundred yards from our office to our car and less than that to our house. We do very little else and so the next chapter, chapter 6, will be devoted to exercise. In this chapter, however, we want to cover more about weight, weight loss, and weight management.

This chapter will briefly look at food and what we eat, but chapter 7 will be strictly about food and nutrition. In this chapter we want to talk more about weight loss and everything that relates to weight loss. The first thing we want to discuss about weight loss is muscle mass. The more muscle you have, the more calories you will burn and the more fat you will burn. This is a simple, well-known fact. Part of the discussion in this chapter and one of the keys to the secret of youth is to be lean. That means getting back to your ideal body weight. That may seem impossible to you. Most of you reading this book are either overweight or obese — actually a third of you are overweight and a third of you are obese, and are looking for answers or an alternative to being overweight and to growing old. We age. We don't have to grow old. Many of you are suffering with illnesses are related to the condition of your body. I want to address this topic with straight talk and deal with this in such a way that you can get back to your ideal body weight and most importantly, that you believe you can.

Since being your ideal weight is one of the keys to youthing and one of the keys to the secret of the fountain of youth, I guarantee you your life will feel different as the youthing process begins. As you actually start to lose weight, build lean muscle mass and get rid of the fat, you will start youthing. You will actually start getting younger. If you weigh 10 pounds less next week, than you did this week you will actually be younger in your metabolic rate and everything else you are doing in life. We absolutely believe weight loss and the balance of fat and muscle is extremely important to your overall youthing process and to how you feel about yourself and everything else in your life. This is the intention of this chapter.

study of martial arts, I didn't have any extra money lying around, with a two-year-old I didn't have extra time, and being 46 years old I was far out of my comfort zone. I felt like an absolute old fool. That choice and the passion with which I engaged in that opportunity, transformed the possibility of my life. Along with my choice, I selected myself to be used by the "warrior spirit." The martial arts are particularly transforming in this aspect. I found that as I engaged in my study of Taekwondo, I developed myself more like an evolutionary leap rather than normal linear development. You know when you are in the "warrior zone," when you are being magnified. When you develop, through consistent right practice, enough skill in your game, the spirit of the game begins to live through you. At this level you will begin to self-express and invent new domains within the game. You become a part of the skill and the evolution of the game. As you invent new domains in your game, the game invents a new you.

We are all small and insignificant. We are all great. It is through passionate engagement in a worthy game, with an intention to serve excellence that we become great. More accurately we become a part of greatness that looks for an opening to express itself into life. Looking again to my own experience, the first element of mastery is to choose to engage in a challenging game. Adults like to be experts, not beginners. It's particularly hard for an adult to be an awkward beginner again. Choosing to play the game is the playing field of mastery.

Another essential element of mastery is recognizing who we are in this existence here on earth. What I want you to get here is the power of your word, because it is integral to living life as a Master. This very principle helped me to complete who I thought I was; that person was really a false or untrue self that developed as a reaction to everything

We cannot talk about weight loss without delving somewhat into what you eat and into how much exercise you get. Although we are going talk about that in more detail in other chapters we will allude to those factors as we move through this chapter. The first thing you have to assess is where you are. Be honest with yourself, here are some convenient charts to help you come to grips with where you are. If you are above the averages in the following chart, you are overweight and if you are more than 30 pounds above the average you are obese. To use this chart simply determine your frame size and then look at your height to determine your appropriate weight. Remember, anything over that range and you are overweight. If you are 30 pounds or more above that weight range you are obese.

Weights in pounds at ages 25-59 based on lowest mortality and according to frame size

WOMEN
Ideal Weight Chart

Height		Frame Size	
Ft. In.	Small	Medium	Large
4'10"	102-111	109-121	118-131
4'11"	103-113	111-123	120-134
5'0"	104-115	113-126	122-137
5'1"	106-118	115-129	125-140
5'2"	108-121	118-132	127-143
5'3"	111-124	121-135	131-147
5'4"	114-127	124-138	134-151
5'5"	117-130	127-141	137-155
5'6"	120-133	130-144	140-159
5'7"	123-136	133-145	143-163
5'8"	126-139	136-150	146-167
5'9"	129-142	139-153	151-173
5'10"	132-145	142-156	152-174
5'11"	135-148	145-159	155-175
6'0"	138-151	148-162	158-176

that happened to me in the past. This may be confusing to read, because it's not the normal way we usually think. It's like a fish turning around to see the water in which it swims. Let's look together at the very air we breathe or what it means to be a human being. There is one thing that distinguishes us from other animals. We have language.

In the Holy Bible, in the first chapter of John, we read, "In the beginning was the word, and the word was with God, and the word was God." In Genesis 1:3 it says, "And God said, "Let there be light: and there was light." This is definitely an example of speaking to create. We could get that God spoke the world into existence, but that is not within our power; we are not gods. Are we not made in His image? If you are agnostic, I may have lost you right here, but consider that to discover your own power to create could be the proof of your own divine heritage. Even if God and religion are offensive to you, stay with me. If you look at yourself deeply and closely enough, you will see that you have always gotten exactly what you want. I can hear you argue against that. "No," you say. "I want love, fame, and fortune and I don't have any of it. I want love, but I have a husband (I write from the perspective of a woman because I am one-you can translate), who is indifferent and cruel. I want fame; but I am a nobody that makes no difference. I want fortune, but I have to work like a dog to make ends meet month after month. I don't have anything I say I want!" What I am talking about is a deeper domain of the power of your word. I say you have had everything to say about the way your life has turned out and that it is because your unconscious word is dominant over your conscious word. It is the discovery of your unconscious word and completing that conversation that gives you the power to create the life you always wanted.

MEN
Ideal Weight Chart

HEIGHT		FRAME SIZE	
FT. IN.	SMALL	MEDIUM	LARGE
5'2"	128-134	131-141	138-150
5'3"	130-136	133-143	140-153
5'4"	132-138	135-145	142-156
5'5"	134-140	137-148	144-160
5'6"	136-142	139-151	146-164
5'7"	138-145	142-154	149-168
5'8"	140-148	145-157	152-172
5'9"	142-151	146-160	155-176
5'10"	144-154	151-163	158-178
5'11"	146-157	154-166	161-184
6'0"	149-160	157-170	164-188
6'1"	152-164	160-174	168-192
6'2"	155-168	165-178	172-197
6'3"	158-172	167-182	176-202
6'4"	162-176	171-187	181-207

As you look at this chart, you can actually see what your ideal weight is. Don't panic and for sure don't get down on yourself. Just try to figure out where you are. It's just a way to say to yourself this is where I am and this is just what's so right now. Someone once said you can't create anything new until you confront where you really are. It's sort of like a map. It's hard to determine how to get to your desired destination until you know where you are. You could come up with a whole list of all the reason in the world why you are not at your ideal weight. Notice that doesn't change anything. That's just the hand you got dealt. It's how you play the hand that makes you a winner. Heredity is almost always thought of first when we're looking to see why we have the body we have. It's actually quite reasonable to think you are the way you are

Consider that you have a life right now that is exactly consistent with your unconscious word and is held as subconscious beliefs and thoughts about yourself, others, and life itself. Are not all human thoughts and beliefs held in language? These words generate your world. They even determine what thoughts you can think...and what you can create. Have you ever heard the saying, "Give me a child until he is six years old and he will be mine forever"? I say that is both true and not true. Even if a human being is systematically and powerfully indoctrinated, there is a lot of evidence to support the theory that even in those circumstances he will naturally exercise freedom of choice and self-determination. Political and cultural indoctrination can be a powerful influence in the domain of what thoughts we can think. Can a Native on a remote island dream of going to the university to study computer design graphics? Well, maybe in today's world, but in the not too distant past that was not a thought a Native would have. That thought would come from a world that didn't exist for him. Most people who are brought up in extreme poverty only perpetuate the same world from which they came. Sometimes, though, someone breaks out and generates a life totally inconsistent with his cultural indoctrination.

One of my favorite books is *Jonathan Livingston Seagull.* If you haven't read it, I highly recommend it. I think it's on the must read list. Jonathan grew up like other gulls. The normal daily existence or culture of the seagull was a continual search for food. The gulls squabbled over tidbits on the beach, or for long hours glided over the water, every so often dive-bombing into the water for the prized catch. They followed the fishing boats screaming and fighting over the discarded scraps. At dusk they perched or nested on the shore, resting for the next day's hunt. All gulls lived and died in this normal expression of "gullness." Jonathan, because of your parents, grandparents, and so on. Your parents may have weighed two or three hundred pounds and no one in your family has ever done anything athletic. You could go on and on and really think there is nothing you can do about it. It's hopeless. As kids we used to cruelly taunt other kids, saying, "When they were handing out good looks in heaven you were last in line and they ran out." You may be saying things like that to yourself now. You might be a woman who has had three or four babies, gained weight you can't seem to lose, and feel hopeless. You might have had an injury and can't exercise. This chart is just a measure of where you are now in relation to your ideal body weight. What I am saying is somehow, some way, if you have the desire there is a way for you to be at your ideal body weight, and will support your vitality and will be the environment for extraordinary wellness.

Let's look at the possible methods of how to achieve your ideal weight. Some of them seem pretty bizarre but we are going to talk about all them. Looking first at some of the extremes, there is liposuction, which surgically sucks the fat out of your body. Some people pay lots and lots of money to do that. It's obviously one possible method of reducing body fat. We don't recommend it. Many people have even died from blood clots or other complications following this type of surgery. It's simply a cosmetic fix, and if your lifestyle doesn't change, eventually your fat and weight will rebound right back to where you were.

Next, there is a procedure called stomach stapling or having surgery that makes your stomach smaller. It helps you to eat less. That's not necessarily healthy either, although we have known people for whom this has been successful and they have lost a lot of weight and body fat. It works for some people...*if* you are willing to take the risks associated with surgery. We are not going to go into that in this book. But these are two extreme measures you actually could use to get your weight down to where you need to be. We want to look at the practical things you can do to start generating weight and fat loss and to start generating the necessary balance between fat and muscle. That is where our discussion is going to begin.

Over the past 25 years a zillion different diet plans and pills have been dumped on us. Most of us could name quite a few. They usually have women's names attached to them, some fad diet of the Hollywood though, dreamed of soaring high in the sky and flying faster than any gull had ever flown. He spent so much time experimenting with breakthrough flying, his parents worried about him.

He was too thin. He didn't spend enough time skimming the waves for fish. The other gulls were beginning to talk and say he was crazy. Jonathan listened only to his inner calling, though, and flew late into the night. Jonathan was bumping up against the limits of the known possible speed and altitude for seagulls. He often stalled and fell into an out-of-control spin, plummeting embarrassingly into the sea. His parents, who urged him to give up such foolishness and do what normal gulls do, sometimes convinced him. He tried to comply. This return to normalcy would last only the shortest time. His mind wasn't on fishing; it was always on figuring out new aeronautical formulas for more speed and altitude. His world was not fight for food but scavenging only enough food so he could fly. He always broke away as soon as he'd scavenged enough sustenance, returning to his training and experiments. On one such day he actually had a breakthrough. He adjusted the angle of his wings and instead of the usual stall as he reached that apparent limit he exploded to a new level of breathtaking speed...with control. He flew higher and faster than a gull had ever flown. Every day he trained and every day he expanded the limit. One day as he practiced in his usually solitary sky, another gull streaked past him at the most amazing speed he'd ever seen. He followed the extraordinary gull. What he found was a place with many gulls just like him. They were all engaged in experimental flight. The gull he'd seen was apparently their teacher. He coached them to unconceivable levels of flight. Jonathan soon became his most accomplished prodigy. We have all heard, "When the student is ready, the Master appears." I say it a little stars, or some new miracle pill advertised on late night TV. After decades of this, people are not only still trying to lose weight... *they're fatter!* Bookstores have hundreds, no, actually thousands of books that talk about weight loss and diet. Fundamentally they are all the same; they go back to the fundamental premise of calories. If you eat fewer calories than you burn, you will lose weight. If you eat about the same number of calories you burn, you'll be at status quo or equilibrium, and if you eat more calories than you burn you'll gain weight. This is common knowledge. While we agree with that to some extent, what is more important about diet is what we eat, as well as how much we eat. Both aspects are important. Improving the quality of the food we eat, or rather improving the density of the nutrition of what we eat, can help us dramatically in how much we eat. We often eat too much because our bodies are craving more nutrition.

Very recently there has been a major breakthrough in weight loss and fat gain that is of interest to everyone. A famous doctor and a few other researchers have come up with this marvelous tool called glycemic indexing. This was actually part of a lengthy study by the National Institute of Health (NIH) in Washington, D.C. They took all the major foods of the world — not groups of food, but actual, individual foods — and determined which foods trigger a higher insulin response from the body and thus generate more body fat. They arranged them in an order and called it the "glycemic index." You can purchase their glycemic index, which identifies the glycemic index of each food and which will stimulate the body to produce and store the most fat. We made that sound so simple. They actually took each individual food and fed it to test animals and eventually to people to see which foods produce more body fat. The testing for each of these foods — and thousands were tested — cost an average of $5,000 each. This was a massive undertaking. It's an extremely beneficial work that took about 20 years. Just think about how far research has come to be able to actually identify a "glycemic index" that tells us which foods to either eat in moderation or to avoid altogether. It even tells us which foods we can eat without worrying about them. We hope you will take the time to visit our website at www.secretsofthefountainofyouth.com; there is a whole section on "glycemic index." It tells how to get a hold of all differently. "When the student is pushing at the limits hard enough and with enough commitment and determination, the Master appears." Jonathan was fulfilled.

One day as Jonathan soared the high skies, he noticed a young gull struggling to fly higher and faster. Jonathan flew down and rocketed past him. The young gull followed. Jonathan began to teach him the new way of the special gulls. Maybe everyone had it in them to be extraordinary. Maybe it was only a matter of time. Maybe it was time to open a flight school on the beach.

Acculturation or even indoctrination cannot kill the divine heritage of a human being's power to invent and create that which has never before existed. It can be accessed in any circumstance, at any age, and at any time in history. It is integral to being a human being. That innate power can be developed much like a talent, if one acknowledges and owns the power of one's own word to create his world. God spoke the creation of the world. "And God said let there be light, and there was light." Our word creates our world. I know stuff happens, but it's what we say it all means that alters our world.

Consider that "the word" has always created your world. Let's look at it. Consider a scenario in which you are five years old. A teacher acknowledges your work and tells you how creative and clever you are. She shows your work to the whole class and says it is the most creative work she has ever seen from any student. You think, "I must be quite extraordinarily creative and clever." You work harder and more and more people tell you how creative and clever you are. People begin to say you must have been born with a rare talent for creativity and cleverness. You are definitely a genius. You grow up and become the most extraordinary, creative, and clever architectural genius the world has ever known.

the literature and all the studies and charts. There is an actual booklet you can carry around with you, so you can look at what you eat and know whether or not it is going to make you fatter.

Basically glycemic indexing is a great tool to help decide which foods support a lean body and which ones produce a body with too much fat. Just like everything else knowing the glycemic index of a food is not an end-all or be-all. It purports some of the worse foods for you are rice and carrots, for example. I notice ice cream is not a high glycemic food and also Snickers candy bars are on their list as a low glycemic index food. I think this science is not perfect because it identifies rice as one of the highest glycemic index foods in the world. If rice were the enemy, all of Asia would be fat because Asians eat rice every day. It isn't necessarily true that rice, even though it has a relatively high glycemic index, is something you can't eat. It's how much you eat. Most Asians eat a very small bowl of rice once a day, whereas in Hawaii they eat massive amounts two or three times a day. Hence most Japanese are skinny and many people in Hawaii are larger, even though they both eat rice daily. It is the quantity that makes the difference. The glycemic index also says ice cream and Snickers bars are low glycemic index, but you can't eat a lot of those or you will get fat and sick. glycemic indexing is a guide to follow, not the "Bible." You don't have to stop eating rice, although I recommend the higher-quality nutrition of natural whole grain or brown rice over refined white rice. Carrots have a higher glycemic index, but consumed in smaller quantities you can still enjoy them and your body can enjoy the quality of the wonderful beta carotene and other phytonutrients found in carrots. Conversely, you also can't go out and eat lower glycemic foods such as ice cream and Snickers bars and think you are going to get skinny. I think the key point here has to do more or less with moderation. We can expand our knowledge of how food is metabolized by the body. Then we can control what we eat and not let eating control us.

What we are trying to do in this book is not be sensational, but tell you the things that work and the things that don't. We don't want you to rush out and try the next fad diet or waste your money and time on things don't work, or could even be dangerous. It is very expensive to try everything that comes along, and demoralizing to fail most

Look at another possible scenario. As far as you can remember, life hurt. You were always hungry and dirty. If you cried, your mother hit you, so you stopped crying. Your mother screamed at you that you were an idiot and would never be able to do anything. When you went to school you were afraid and didn't speak to anyone. You were hungry and often fell asleep. You never knew what the teacher was talking about. You failed. Other kids called you stupid and then you didn't try at all. Finally you got old enough you quit school. It must be the gene pool you got. You came from a long line of idiots. All you could find to do was wash dishes. Your boss yelled at you for being stupid. You couldn't even wash dishes right. What else could you expect of someone so stupid?

These are of course exaggerations and still opposite ends of the same domain. When you are young, your parents and teachers are the "gods" of your world. In order for their word to have power to create your world, though, you must add your word to what they said. You have to believe them and give your agreement to it. X happened to you and you said it meant Y about you, them, and life, and that was it on you. That's known as "false cause." Sometimes you did not accept what they said. You decided to be the opposite of what they said. It's still the same domain, one of reaction to what happened. You didn't own the power of your own word. You think they did it to you. You did it to yourself, because you gave your power to their word. That is OK, YOU were little. There was nothing else you could have done. It's the human condition. Now, though, you won't hear, see, or think anything not consistent with that reality. It's all held in place with your word, but mostly as an unconscious reaction.

The first element of mastery is to accept responsibility for the power of your own word. It is to be aware of, accept, of the time. We want you to win and do it affordably. We will share what we have learned and what has worked for us. We believe weight management has many facets and we still don't know one certain way will work for everyone. We will present the things we have tried and proven, recognizing each person is different and responds differently to the same method. There are, though, many eating plans or diets that are just contrary to common sense. If you eat ice cream and Snickers bars you may not gain as much weight as eating an equal amount of rice but you will probably die of a heart attack. The glycemic index model doesn't account for what sugar, ice cream, Snickers bars, and the like does to your cholesterol levels or blood pressure. We do think glycemic indexing is an important guideline to follow or we wouldn't be mentioning it here.

Most popular diets are called fad diets for a reason. It's because everybody goes through cycles where they hear about a diet working and are encouraged that there has finally been a breakthrough that may work for them. They buy the book, buy the foods and follow it religiously. They might even lose some weight. Most people eventually stall with the weight loss and finally give up. What's worse is after going off the diet they usually end up fatter than they were when they started. They sink back into hopelessness and frustration until they hear about the next miracle breakthrough diet, that thousands are losing the weight they never could before. They buy the book. Sounds familiar doesn't it.

The cycle seems to be the same each time. First the "authorities" on the subject say, "Don't eat any fat because that is why the Americans are fat." Later they found out that is not really why Americans are obese. When you eat fat it doesn't affect your weight as much as it affects your heart. You can die of a heart attack or stroke if you eat too much fat. glycemic indexing is getting more to the core of things. The glycemic index of food actually tells us which foods generate more body fat. I also think what you'll find more important than any other factor — and everybody who is fat or overweight knows this — it is how much you eat that really determines how fat or skinny you are. I think even skinny people like me have been through the cycle. I know at one point when I was in college I just ate too much and didn't exercise enough and ballooned up to 175 pounds within a relatively short period of time. I and embrace that you have had your life be exactly the way it is. Can you see that if you could take responsibility for giving the power of your word away, you could take that power back and speak a new world into existence? It is the ultimate sense of responsibility. I learned this for myself in that life-altering course in which I participated in 1979. The course is still offered in many cities around the world. It is presented by Landmark Education Corporation and is called the Landmark Forum. This experience was a line of demarcation in my life. You can check our website for information on their courses. I highly recommend it. Do you remember how I said that into each life come tiny cubic centimeters of chance that if taken, can alter the very possibility of your life? Well this was one of those cubic centimeters of chance for me. The course cost $350 and that was a huge sacrifice for me back then. I drove eight hours to another city to do it. I was answering an inner call, though, not the outer voice of what was reasonable and comfortable at the time.

In the Landmark Forum I learned about the power of my word. At the time I was divorced and raising my son on my own. I had a fairly successful business of my own and loved my son, who gave me so much joy and filled my life. I wanted a husband and loving partner, but I was not unhappy. I had been divorced for about 10 years. A loving relationship with a man just seemed to not be my lot in life. In the Landmark Forum I saw how that would turn out to be my lot in life, given my relationship with my father and mother. I saw that I put together a belief system of the way it was in families and darned if it didn't always turn out that way. After the Forum I met my husband, Ron Zeller, with whom I am still madly in love with after 25 years of marriage. He's the most exciting, empowering, loving partner I could never have imagined for myself am 5 feet 7 inches tall by the way, which, with my small frame, pushed me into the obese category for several months. Someday they may have a pill that does it all, and as a matter of fact, we are going to talk about an unbelievable product in a later chapter that is a magic bullet of sorts. It comes as close to that end-all-be-all pill as anything I can think of that would help you to control and maintain your weight.

What comes next is the most important part of this chapter. What we have to say about weight loss is it is work. It is a four-letter word. W-O-R-K. What this means is when you work for something — and I don't care what it is — somehow it seems worth it. It doesn't matter whether it's losing weight or doing exercise, whether it's something you want in life or some accomplishment; it somehow seems worth it in the end. The achievement is worth more than the nice physique or the slim body. It's the work that got you there that is really worth the effort. That is the most satisfying, the most important part of life, and the most important message in this chapter. The adage, "Anything worth having is worth working for" is the motto by which we live. We live it and we know it's true. No matter where you are, no matter how overweight you are, no matter how sick you are, if you want it bad enough and work hard enough, you can have it. That is the message of this entire book. That is at the core of the secret of the fountain of youth — you must have the desire to transform your life and to be willing to work for it to have everything you want in life. The wellness you want, the health you want, the ability to do anything you want, any time you want to have and truly do have a fountain of youth flowing through you is what this book is all about. We believe that the core and the very heart of all of those things is the word — work. We believe if you want it bad enough you will work for it. It is in the doing of the work that the power of the secrets of the fountain of youth is realized in your life.

As we look at weight loss we need to look at what it actually takes to do it. First you have to be wise in what you eat. Over the years, we know that the government has changed their main food groups and their recommendations of what you should eat and how much. What we found as a general rule of thumb is that if it is living, if it has enzymes in it, if it is a fruit or a vegetable, it is good for you. Even fruits have sugar and if you eat too much fruit you'll probably be overweight or obese.

— until I accepted responsibility for always having it turn out exactly the way it always had before. If I always had it be that I was unloved and alone then I had the power to now choose to be loved and belong. I chose love. I am loved and fulfilled. In fact, my life is so filled with love and beauty I can hardly believe it.

My experience in the Landmark Forum left me with an entirely new world of freedom and power. I began to paint a new life on my blank canvas. It's very interesting that the life partner I yearned for showed up soon after. The curse had been lifted. The empowering part was that I was the one who cursed me to a life without love. As soon as I recognized that, I freed myself. To tell you the truth, I didn't consciously choose living a path of mastery. I just kept engaging in the opportunities life offered, with courage and commitment, and the path of mastery chose me. The Landmark Forum gave me a new and natural full self-expression. Here I was at an age when most people think it's too late to know what you want to be when you grow up. I naturally wanted to experience more of life. I went back to school. I studied music. I got married. I had another child. After my second child at the age of 44, I wanted to get back in shape so I enrolled in Taekwondo classes.

Now we have gotten back the power of our own word and our power to create. What will you create? Anything you want. That's right. You can create anything you want and are willing to do the work to realize. Oh, the bad news. More work. No! That is the good news! Passionate engagement in the work you consciously choose for yourself gives extraordinary life to you. It's the literal fountain of youth. Oh, you don't know what you want? Just look at those times of your life when you felt most alive and passionate. Choose there. For me, I love sports. Some

It's just like eating too much cake, as silly as that sounds. Carrots have a high glycemic index, as silly as that sounds, but, they have an very high amount of beta-carotene and other phytonutrients, which are incredibly good for you. It's not like you could O.D. on carrots, and we don't think because they are high glycemic indexed that you are going to get fat eating carrots or that you should stop eating them either. If you eat carrots, you don't want to eat too many of them. Educate yourself about foods that are high glycemic index and those that are low glycemic. You need to know about all foods and their glycemic index rating and adjust your intake. Cut back on high glycemic foods and step up your intake of low glycemic foods. Go to www.secretsofthefountainofyouth.com and get all the information you need. It is that simple. Youthing begins with information.

News Flash! Xocai Chocolate is Here!

Weight Loss as simple as eating chocolate every day? Is that possible?

Our new weight loss program will alter your life. We have discovered a new chocolate product that proves to be one of the most intriguing new keys to losing weight.

There is a new chocolate out on the market that is made from unprocessed cacao and combined with Ascai berries to create the most astounding chocolate nuggets you have ever eaten. These products are more than healthy, they provide you with almost 4,000 ORAC's of antioxidants per nugget. They are part of a new health and weight loss program that is changing the face of diet and nutrition. It is so amazing. It is called Xocai Chocolate and all the latest information and how to get the product is now available on our website. We actually have a whole entire weight loss program centered around eating Chocolate! Visit our website right now to get the latest information on this and related revolutionary chocolate products. Can you imagine getting healthy and losing weight by eating delicious chocolate. Visit our website!

Vegetarian

Vegetarian? There are a lot of vegans in the world today. Vegans are people who don't eat anything other than vegetables, beans, nuts, seeds, and grains. They don't eat dairy products of any kind. They don't eat of the best times of my life were engaged in sports and competitions. Wouldn't you think at 46 years old I might think it was too late for a choice in that domain? Again, I say, listen to your inner calling, not the voice of what is normal and reasonable. When I first started my classes in Taekwondo I was looking to recover my physical strength and a body I could again be satisfied with. Although when I got the invitation to go to a competition, I didn't say, "Oh no, I'm too old for that." I said, "OK, I'll do it." Just like the song that says when you get the invitation to dance in life I hope you'll dance. Accepting those invitations to dance brought me to a life engaged in mastery. Just choose a game. It's the playing field of mastery. I recommend you choose a game with competition or performance. It's the measure of mastery. Don't worry about choosing the wrong game. It's not sealed in stone. If you see it's not your game, choose another one. Again, the game of martial arts is particularly suited to the game of mastery. It's also great to be able to defend yourself in today's world…but concert piano will do.

Webster's dictionary defines a game as any form of play, amusement, or sport involving competition and with rules; a business or job, especially involving risk. Games have a structure, rules, objectives, and winning and losing. That is the risk aspect. Again I like the Martial Arts game. It engages the mind, body, and the spirit. Competition and performance are definitely available, which stimulates a challenge to higher levels of development. Competition clearly exposes our weaknesses to us. This is a good thing. It requires you to work on your weakness in order to advance and grow. Most people avoid anything that would expose their weakness, even to themselves, much less in public. Winning in public is quite easy, but if you compete you will fail in public. Most don't want the risk of looking bad in eggs. They don't eat animal flesh or even fish. By the way, I just found out that Mary Louise has started a new Vegan nutritional program and loves it. She says her digestive tract feels much better and her energy is way up. Look at her chapters to see what she is doing!

We don't believe it is essential to be only vegan in your nutrition to be your best. It's one possible way. Experiment and see how your body responds. If you will, we applaud you. Many people think we should be herbivores and not carnivores or omnivores. Again our philosophy and experience is moderation in all things. Yet, science proves that if you want to lose weight and fat, eat more and more vegetables as close to their natural state as you can, moderate quantities of whole grains, beans, nuts, and seeds and less of everything else. Eat less sugary foods, less fatty foods, less meaty foods and less of everything you already know is bad for you. We mentioned drinks in the last chapter, but just a quick reminder, avoid alcohol, coffee, soda pop, and fruit juices, especially orange juice.

You need to make a decision in your life. Are your health, well-being and youthing journey worth the changes you are going to make? Some people would rather die than change their diet and they usually do. So, it's time for you to make a decision. That is another of the secrets of the fountain of youth; it is your life, so what are you going to do with it?

The quality of your life is not necessarily about your heredity. It is not about the people around you or your environment or anything else. It is, however, about you and the decisions you make about life. Those decisions are about what and how much to eat and drink in order to have a different quality of life; a transformed life. The biggest part of that decision is being at your ideal weight and fat level. If you are willing to commit to doing that, no matter how much you weigh, no matter what percent your body fat is, no matter where you are at in your life, no matter how depressed you are etcetera, etcetera, etcetera. If you are willing to just say to yourself, "Okay, I'm ready to live life differently. I'm ready to educate myself about this body I have and be responsible for being in the best shape of my life." Then you are on the road to earning and having all the secrets of the fountain of youth. The first thing you do is change your diet. Start eating less. Mature adults do not need as many calories to maintain as children need to grow. As front of others. They will only participate in competitions from their comfortable seat in the stands. They might want the fun of playing, but they are not willing to risk losing. One of the biggest obstacles even for those on the playing field is holding back for fear of looking bad. In order to win you have to risk losing, even looking ridiculously bad. That has got to be OK with you. The path of mastery includes losing as well as winning. They are equally important and an integral part of the path. When it's OK to lose you are free to win.

Your actions will obviously be perfectly coordinated with your intention. Either you will have an intention to avoid looking bad in front of others and your playing will be perfectly coordinated to that intention, or you will have an intention to win and your playing will perfectly reflect that intention. Part of mastery is identifying your true intention. It's often not what you say. You have to look deeper. Your true intention is always plainly evident during the playing of the game. The only way I have been able to identify my true intention before the game is to ask myself the question, "If I get to the end of the game today and I have lost, what is the reason why I lost that I'll tell myself then?" Very often my underlying considerations bubble up to the surface and I can make a conscious choice to just play the game to win. I'm reminded of a time when one of my outstanding students was competing for the first time after an injury from a previous competition. She is six feet two inches tall and a national champion. Her opponent was five feet six inches, slow, and not experienced. My student was beaten badly. She didn't throw a kick until the last 30 seconds of the last round. It was one of the most frustrating coaching experiences I've ever had. After the fight I asked her what happened, why she didn't do one thing she knew to do or what I coached her to do. She blurted out through you learn, incorporate the changes one by one. Mary Louise told me she had to first commit to eating something raw at every meal. She then incorporated eating only whole grains and decreased the amount of grains in her diet. She only lightly steamed or stir-fried her vegetables. She eliminated fruit juices to decrease her sugar intake. She eats no more than two or three pieces of fruit a day and eats them early in the day to use the energy for training. She eats them alone, because they require different digestive enzymes than do other food and require less time to digest. She decreased starchy carbohydrates, (grains, potatoes, corn, beans, etc.), and ate all the fibrous carbohydrates (mostly green and other color vegetables) she wanted. She was encouraged to change her diet to totally vegan 10 years before she was ready to do it. When she hit 60, she knew her body would do better and continue to thrive if she changed her diet to totally vegan. She then took that next step. When you make one adjustment, the next step just shows up. You'll feel drawn or pulled to do what's next. Don't stress over it. Every now and then eat something you really love just for the taste…like death by chocolate! Just make sure it's more *then* than *now*. Mary Louise declares a "have anything I want day" once a month. Even then her commitment to health rules the day and she uses the "moderation in all things" guideline. There's one thing we don't do, even in moderation, and that is consume that which poisons our bodies, such as alcohol, smoking, drugs, sodas, and even any carbonated drink. Go to our website www.secretsofthefountaionofyouth.com and get started on the weight loss program immediately.

We mentioned Europe and we mentioned Japan and why the people there are skinny. My wife and I recently visited Paris and one of the things we noticed is that we walked a lot. We rode a bus and we rode a train here and there, but then we walked everywhere, as do the French people. Once you get off the bus or train there, you have to walk quite a distance to get where you want to go. In America we don't do that a lot; we mainly ride around in our cars. I also spent several years in Japan. People there walk everywhere too, even more. They walk from work to lunch. Like the French, the Japanese people walk from train stations and bus stations to their homes, to their work places, they walk to visit.

her tears that she thought she was very brave to go back into the ring after being injured the time before. She was afraid of being hurt again. Her playing was perfectly coordinated with what was going on in her head. She was brave enough to go into the ring and did exactly what she needed to do not to get hurt again. Winning wasn't on her list that day. If we had a conversation to reveal her underlying concerns and fears we might have been able to defuse them before her competition. We certainly do it now before every competition.

When you are aware of your fears you can choose not to act them out. That is called having or owning your fear. The other way, it has you and acts you out like a puppet on a string. Without awareness of your fear, the competition becomes the medium of expression of your fear. Just like subconscious conversations being dominant over your conscious ones, subconscious intentions have dominance over conscious ones. One of the elements of mastery is developing your ability to uncover subconscious intentions and take responsibility for consciously creating your intention with regard to your game or anything else, for that matter.

Games don't automatically produce mastery. It depends on you. Some people want to develop skill to win and be the best. Even at that level, you can't escape some of the lessons and rewards of mastery. In order to be on the path of mastery you must desire and seek mastery. It has a cost. You can be a great technician without integrity, but you can't play the game of mastery without it. You cannot have the power of your word without integrity. If, at the basic sense of integrity, you don't tell the truth or keep your word, will your word have much power? No! I guess I have a fundamental belief that all humans are perfect and Godlike. I think we have a barrier to knowing our

Exercise is one of the key elements that is missing from our modern-day lives. Although we are going to spend the next chapter talking about exercise and the benefits, we will mention here that the way you burn calories is by exercising...*moving your body!* Exercise raises your metabolic rate. That's what keeps your body lean, by the way. A higher metabolic rate generates an efficient calorie burning system, which is otherwise known as a healthy, efficient, good-looking body. We believe in exercise and we believe in a lot of it. We believe most people should get at least an hour a day of exercise. Make the time. Watch less TV. Get to bed earlier. Get up earlier. Bump it up to two hours. Your body will love it. It remembers muscle tone. It'll strengthen. It'll drop the fat. It will thrive in health and well-being. Move it or lose it!

Regarding fat to lean muscle ratio — when you build lean muscle mass you obviously begin burning more calories and when you burn more calories your fat goes away and when your fat goes away you usually lose weight. We believe the loss of body fat is a key issue to health, youthing, and vitality. Begin by measuring your body. Measure every month of your program and if you do you are on the right track, those measurements will reflect the loss of fat and thus size. Just a quick note that everyone already knows, women have slightly higher body fat percentages than men. If you are a good male athlete you could get down to 5 to 13 percent if you want to have extraordinary health and a body you'll love. If women want the same result they'll have it with a body fat of 20 percent as represented in the chart below. Many Americans have a body fat of 40 percent or more, which puts them at a high risk for heart disease and stroke. Maybe that explains why we have the highest rate of heart disease and stroke in the modern world.

Body Fat Percentages

	Women	Men
Minimum amount to stay alive (this is not a healthy range for most people)	10-12 percent	2-4 percent
Lean – mainly found in athletes (perfect for serious athletes)	14-20 percent	5-13 percent
Ideal – indicates a good level of fitness (a healthy range for most people to shoot for)	21-24 percent	14-17 percent

godliness and the perfection of our true nature. We often lose our way doing all sorts of things that are inconsistent with our godliness, such as lying, stealing, and even worse. I think at some level we always know we are doing wrong and won't allow ourselves to succeed or win. We might temporarily, but not in the long run. Even if we are dull enough to allow success, we often don't experience the satisfaction that should go with it. In martial arts, integrity is acknowledged as a basic tenet, yet I've watched people teach that tenet while they themselves are lying, cheating, and stealing. I've never seen them last. It's as if the spirit of martial arts eventually spits them out.

If the basic level of integrity is telling the truth and being honest in all your dealings in this world, the ultimate sense of integrity is being true to your ideals and standards. To do that you have to examine yourself deeply to uncover what those ideals and standards are. When my husband and I got married, you would have thought our marriage was a merger of two big companies. We examined our expectations of each other, our standards and ideals for relationship and family, and the commitment and promises we were both bringing to the marriage. We wrote these out clearly in a contract form. That may seem obsessive, compulsive to some people, but it really served us to see if we had enough in common to make our marriage work. Most people don't find out how little they have in common in some very important areas until after they are already married.

One of my mentors and someone I admire in this world is Grand Master Jhoon Rhee. He is recognized as being the "father of Taekwondo" in the United States. He is not only incredibly skilled in Taekwondo but he embodies the true spirit of master. He is a man of integrity, passion, commitment, intention, empowerment,

Average – not that great (many people fall somewhere in this category)	25-31 percent	18-25 percent
Overweight or obese – *over 50% of the country* *(you don't want to be in this category)*	32 percent plus	25 percent plus

More on fat!!! Another way you can track your body fat is through scales; there are new body weight scales out on the market that are able to give you your weight and your body fat. It's not as accurate as more scientific methods, but it is a good tracking system. We believe anything you do in the form of exercise will help that ratio. The more lean muscle mass you build, the better off you are as far as cardiovascular health. The more lean muscle mass, the higher the metabolism, and the more fat you will burn. The more fat you burn, the better your health and the better you look. Just do it! The results are worth it. You'll not only live to enjoy life longer, you'll enjoy it more. At some point there is a trade-off, though. Some people workout to extremes. This is unhealthy as well. There is a point of negative return. They are destroying more than they build. Many world-class competitive athletes die an early death as well, as a result of too much and too intense exercise. The ideal human body is more along the lines of somebody who's aerobically fit, the best example are swimmers, runners or bikers. Everyone has seen them; they don't have great big bulging muscles. They are slim and trim and they are at their ideal body weight, with an ideal lean-muscle-to-fat ratio. We are once again talking about moderation in all things. The ideal body is not a six-pack and great big bulging arms and legs. The ideal body is just having great aerobic conditioning and a thriving, healthy toned, beautiful body. You can be fit and look good, but you don't have to have bulging muscles. Many athletes who have those big, bulging muscles are paying a price with their health. Many even destroy their health with steroids to get that look. It becomes a destructive addiction. To sum it up, what you need is to get aerobically fit and watch what and how much you eat and drink. That is it. It's really not rocket science. Youthing is maintaining your ideal weight, body fat and muscle mass.

Go to Chapter 9 for the whole story on what to eat and drink and what supplements to take if you want the quality of your life to go off the charts!

courage, perseverance, invention, and discipline. He has used Taekwondo to make a huge difference in the lives of thousands of other people. He taught me the principle that man is elementally perfect. It just makes sense to me. He says, "Were we not told to be perfect as God is perfect? We are made in his image. We are perfect as God is perfect. If we make a mistake, we just have to clean it up. When we clean it up, we are perfect again." That is what integrity is to me. It's being honest enough to own up to our mistakes and cleaning them up. When we clean up our mistakes, we are whole and complete again. Integrity is being whole and complete. Without cleaning up our mistakes we eventually lose sight of our real selves. It's like having a beautiful new house and everyday ripping a board off until there are gaping holes in the structure. The integrity of the house is so damaged that it can't protect us from the storms of the environment. At any point we could take on rebuilding this house, restoring its original integrity. It would again serve its function of protecting us from the buffetings of the world. So it is with our integrity. We can't have the blessings of living the life of mastery without taking on integrity.

Games require skill and strategy. In order to learn the skills and strategies needed to win, we must have a teacher, a mentor, who has the experience to pass them on to us. If you want to be the best, seek the best teacher. The best teacher will have the best students. Those will be the students winning the competitions. It's not only about winning, it's about the right spirit. All you have to do is pay attention to discover if a teacher has the correct spirit. You will know if you are being empowered. My first Master Instructor, Master William Kim in Vallejo, California, taught in a very positive manner. He wasn't patronizing; he always looked for ways to speak encouragement. He

CHAPTER 6

EXERCISE AND FINDING A TRAINER

The Key To A Balanced Life

Perhaps more than any other chapter, this chapter is designed to help you put it all together; using the brain, the word, the declarations, the health and wellness, the weight management and optimum body fat to muscle ratio and putting them all together into the new and "real" you. Chapters 5 and 6 are actually understood and used in tandem. Without losing some weight and body fat, perhaps you can't even think of exercise, and without exercise you are not going to lose any weight or body fat. Basically this chapter is about how to do and have both. This chapter will help you see how to enter into the empowering world of exercise and how to lose weight and body fat. The result of the next two chapters will show you how to have a body you love.

Many pills have been developed to boost the metabolism to try to stimulate your body to lose weight unnaturally. We unequivocally state that we believe that exercise, which involves passionate and focused body movement, not magic pills, is the key to life. Moving the body vigorously with full involvement of the mind and being, is therefore one of the acknowledged some breakthrough in your technique or effort as he gave you a correction to take you to the next level. Master Kim was an experienced competitor who knew firsthand what every competitor had to go through on the path of mastery. You have to be willing to be a beginner before you can be a master. Find a true master to show you the way. Don't settle for a teacher just because he is close to where you live. Do some research. Get recommendations and observe several instructors while they teach. Get on the Internet and see if they really are who they say they are. I know someone in my city who teaches martial arts, who claims to have been on the Korean national team and advertises that he is a 12th-degree black belt. First of all, there are no 12^{th-}degree black belts. The ranking system goes only to 10th. On the Internet he is not listed as ever having been a Korean national team member. He's only a fourth degree black belt. Buyers beware. You will recognize a good teacher when you see one. Keep looking until you do. You'll go nowhere with a mediocre teacher. There are many teachers and few masters. Find the master if you want to be one.

Now you have found a good teacher. Do you know what the matching bookend to a good teacher is? That's right…it's being a good student. Begin. Beginnings are the seeds that grow mastery. Sometimes beginnings don't point in the direction of extraordinary anything. Perhaps the real reason to always be a beginner at something is to cultivate humility and an open spirit. Being an awkward beginner is a humility machine. If you can truly allow yourself to experience the awkwardness of a beginner, the humility it produces keeps the spirit pliable. I also hear you'll never get Alzheimer's if you are always engaged in learning something new. The more you say "yes" to opportunity and participate with passion, the more opportunities will seek you out and greatest secrets of the fountain of youth and one of the keys that literally unlocks the door to youthing. We are going to present exercise in such a way that you will discover for yourself that using all these things in concert...the brain, the power of the word, the declarations, the goals, the weight management and the optimum body-fat-to-lean-muscle ratio, will literally unlock the door to youthing for you personally.

We have noticed, and maybe you have too, people who exercise typically appear vital, youthful, and powerful; they appear to just have their lives together. We want to emphasize our fundamental philosophy right up front that we believe every person who seeks access to the secrets of an extraordinary life, especially about youthing, must exercise. Remember exercise is only the code word for play involving exuberant whole body movement. It may have been many years since you have even thought about exercise. You might think, well, I weigh three hundred pounds and I'm five foot eight — how in the world am I going to start exercising? I am going to play? Me play? You, yes you, just the way you are right now — you can exercise (play) and you will immediately begin to reverse the drift that got you the body you now have, the body is just struggling to maintain life. Moving your body every day will transform it and give it life — real vitality. You must move your body! Choose life — move! All living things move! Vigorous play and movement will generate and maintain your ideal body weight and fat content. Move it! Get a game and play. Play and move every day. If you want to build a business to a brilliant success, what do you do? There's a right action to produce every brilliant result — Right? If you want a brilliantly successful business, or a brilliantly successful marriage, or a brilliantly successful anything...you take consistent right action. There are fundamentals to every success. One of the most basic fundamentals to a successful business would be to show up for work every day. You would be smart to have experts/mentors guiding and coaching you to take right action every day. The daily practice of right action must produce success in any domain of life! It is, of course, the same with your body. Mary Louise and I are offering ourselves up as experts, coaches, mentors, that is business consultants for your body work. We've walked this path for many years...45 combined years to be exact. We've researched, work through you. When you magnify an opportunity with your effort, it then magnifies you. Engage fully. Go to class. Going to your classes is a priority with you. Rather than worrying about how you look, you focus on the technique and try your best to do the skill your teacher modeled for you. My Master Kim used to always say, "Watching is halfway there." You might do a technique a thousand times looking like a jerk, but then one day you can do it and then you look so cool. At every class be hungry for learning with the intensity of a tiger looking for lunch.

The only path to mastering a skill is practice. A good student engages in practice with consistency and a passion for excellence. You know you have the right game if you love your practice. I am always nurtured and enlivened in my practice. I don't always want to practice. The only thing I really always want to do is watch a LifeTime For Women "he-done-her-wrong" movie or American Idol. Once I get past the "I don't want to," I am very enlivened by my practice. Full engagement in consistent practice is truly the fountain of youth. I'm afraid to stop. I think I would turn into an old grandma right away. Many people guess me to be 20 years younger than I really am. That is one of the rewards of passionate practice and continuing to dream "the impossible dream." Mastery of technique is accessed through practice. In the mastery of technique you also learn to master yourself and life itself.

Miyamoto Musashi, one of the most famous ancient Japanese samurai warriors, who won over 70 battles to the death, wrote about his practice to develop his skill with his sword. He tried hundreds of times to slice a small pea on the ground with one swift strike of his sword. He missed more often than not. Every day he practiced hundreds of times. One day, instead of just increasing his average of success, he had a breakthrough. He said that instead of the sought out experts all over the world to coach us. We've used our own bodies as the "guinea pigs," the "lab rats" and the results are in…right action in the area of nutrition and exercise will produce a body that will be vitally strong and alive ***all*** the days of your life. You can have vibrant health and beauty of youth ***all*** the days of your life. There is no age at which you ***must*** lose your vitality. It's a mean myth — a very bad joke. It's a test to see if you will own your own power to create in the face of an all-pervasive lie. The body is not the limit! You just got sucked into the lie!

Typically, what happens is when people who are overweight or obese actually begin an exercise program they do too much and either get injured or have a serious health stress reaction and then stop exercising. I don't know how many people I know that have a knee problem, a back problem or this problem, or that problem. An important part of our coaching is to coach you through a safe initial phase. You must first get to the level where you can exercise. We will help you get to that point. We will show you the way.

Obviously you need to talk to a physician before you begin. That is always an essential recommendation before you get started. A physician can assess your current level of health and fitness. Remember, you can't create until you know where you are. If you have any illness or you are so obese that you can't do certain things without putting your life at risk then you need to start by getting your body well. There are some mild exercises, though, that can be done from a chair. You want to begin being informed by and responsible for your current state of physical health and fitness. The medical profession can tell you what you can or cannot do. That having been said, we encourage you to engage in our coaching program to turn around, expand, and even explode your vitality and physical well-being — your physical prowess. Can you begin to think of yourself as having physical prowess? Take it on. Begin right where you are. Your first goal may be to get to the point where you can at least begin exercising. If you follow the protocol of this program, no matter what state of health you are now in, no matter what you have been doing up to now, no matter what you have always believed about your physical body and what you are capable of doing, your inevitable result will be having a body you love and a body that pea looking so small as it always did, it suddenly appeared to be the size of an orange. He sliced it exactly in two every time after that. His perception and skill expanded to a whole new dimension. The one who wins is almost always the one who has practiced the most. Mastery naturally requires the daily discipline of passionate right practice. We already know it's right practice because you have the right teacher. When you bring passion to the practice it becomes passionate practice. In addition to greater skill, love of excellence in all things is a natural by-product of passionate practice. Practice transforms the player into the master.

Choose big goals, even impossible goals. Just engage in them as if they are possible. When Roger Bannister broke the 4-minute mile, it was considered to be humanly impossible. He trained like it was possible. When he did it, about seven other people did it the same year. It's OK to dream the impossible dream. Be a dreamer/doer. The bigger the goal, the more you grow. I am such a big proponent of competition because competition can teach you what you will never get from training and practice alone. There is learning and strengthening in both winning and losing, so you can't lose. It is in competition that your skill and strategy make exponential "mega-leaps." Integral to a game is having a measure of progress. Competition or performance definitely is a measure of progress. There is also no way to get experience in your game other than to get experience in your game. It takes courage to compete. Just like putting weight on a muscle, it grows. I have always been nervous when I compete. That is not a feeling I like. I do it anyway. That is how to grow courage without having a crisis in your life. I call it "planned adversity." The times I've won have built my self-esteem and confidence. The acknowledgment of winning has given me the acknowledgment I never had growing up. My accomplishments in both rank and will support you in doing your dreams, goals, and visions all the days of your life. We promise you! Your first task may be to just believe you can get to a point where you can exercise in some way.

This part of the book is dedicated to how to determine your ideal level of exercise and identifying what that exercise should be. If you are extremely overweight and if your doctor has said it's okay to do some mild form of exercise, we can direct you. If not you'll just have to wait until you get to that point by following the nutritional and supplemental protocols we have recommended. When you get to the point where you can at least start mild to moderate exercise, this is the starting point of the rest of this chapter.

What type of exercise you do is a very individual choice. Some people like running, some like swimming, others like biking, walking, Tai Chi, Yoga, ballroom dance, modern dance, and most women like some form of aerobics. Mary Louise fell in love with the mind, whole-body and spirit-strengthening aspect of the martial arts. Acquiring self-defense skills in today's world is also not a bad idea. You must remember for our purpose here we are talking about aerobic exercise and not weight training or bodybuilding. The human body needs aerobic (oxygen) conditioning. It is essential to life. Personally I prefer running, and for many individuals it is the ideal form of exercise. However, there are many other games, self-expressions, and ways to get exercise that will fulfill the body's essential need for aerobic conditioning and muscle tone. Choose a game. If you discover you don't enjoy it — no problem. Choose another one. Just choose one and engage in it every day.

Exercise that does not cause too much stress is the ideal form for the beginner. If you are overweight or you have to start at some level other than running you could do things are less stressful than running. Those could be swimming, walking, biking or some other form of exercise on equipment of some sort. That is where you need to begin. Running is excellent. Biking is excellent. Aerobics is excellent. Swimming is excellent. The only problem with swimming is that you need to find a pool to go to regularly. The only problem with biking is you must be able to afford a bike. The great advantage of running or walking is that all you need is a pair of running shoes, something to wear and head out the door. Let's look over this whole gamut of exercising. You titles inspire me to desire greater levels of perfection and excellence. Losing has refined me. It has expanded my capacity to accept disappointment with the grace of an adult instead of with the grief of a child. Accepting the full experience of failure again grows the courage to keep generating my goal over time and through all experience. Losing truly helped me grow up. Losing also inspires me to seek higher levels of perfection and excellence. Losing in particular shows me my weaknesses. Most people don't play because they are not willing to experience losing. I suspect you can't fully experience big wins in life if you avoid the risk of losing. Given that playing the game, win or lose, is the source of mastery, I say play!

An extraordinary person is an ordinary person who engages fully in an extraordinary project or game. Aren't those the times that you also feel most alive? Do research about your project or game. Learn from others all you can. I've read books, watched videos and participated in training and workshops all over the world. I never had enough money to do all of that; I just always found a way. When you have engaged in your game in this way long enough, it will get a life of its own, leading you to more opportunities. You begin to invent new possibilities in your game, making personal contributions to the game. As you invent new possibilities with your game, your game invents new possibilities of you.

Seeking out and engaging in a new experience, consistency, perseverance, diligence, and discipline are the ingredients of mastery. I've taught a lot of talented people. Talent doesn't make a champion or one who will eventually make a difference for others. Give me a willing, passionate, consistent, persevering hard worker every time.

Dare to choose a game, seek a mentor or coach, engage with passion in a daily discipline of training and can find out many things about exercise by going to the bookstore or looking online. No matter what form of exercise you want to get into, we suggest that you become an expert even before you start. If you can, get a coach, even a friend, who has more experience than you.

Once you begin exercising, realize that if you can work up to one hour per day of exercise that is what we have discovered is one of the keys to the secret of this fountain of youth. A minimum of one hour of exercise a day is the optimal point at which your body functions; it is inherent in the human body's design. We have tried everything. We have tried less than that, and we have tried more than that and nothing else works better than one hour a day. Let's just say experience has taught us this. You need to exercise between one and one and one half hours daily. Mary Louise and I rest one day a week. We take Sundays off.

I've been running for about 34 years. When I started out I was not just overweight. I was obese. Just to give you my personal statistics I was 5 feet 7 inches and I weighed 175 pounds That would be obese because my ideal weight is about 140 pounds I was 35 pounds overweight and even though I didn't look obese, I was. Just graduating from college, I hadn't run a step other than taking one PE class when I was a freshman. I was forced to run to get a good grade in PE and so in three and a half years, since that elementary PE class, I hadn't run a step. I was eating everything I wanted anytime and anyplace I wanted. I ate. I slept. I drank soda, and everything else you can imagine. I had two children. I was in school. I worked 35 hours a week. All I did was sleep a few hours every day, eat, do my homework, eat, go to my morning classes, eat, go to my afternoon classes, eat and then I'd go to work and then I'd eat and I'd eat again before I went to bed. I didn't do one single ounce of exercise for over three years.

It had gotten to a point where my face was plump and round and my waist size was up to 38 inches. Basically I was in the worst shape of my life. One day, after I had gotten out of school, I went skiing with my uncle and I felt like I might just die after skiing only a few hours. I decided then and there to do something about it — because I loved skiing! I was young. I wasn't ready to give up skiing. I decided I would start running. The next morning I started my new running program and I honestly could jog only one lap around the track. Even then, I practice, create the possibility of greatness by creating an impossible dream, challenge yourself with competition and performance, measure your progress with advancement in your goals, and then really set yourself up for growth by teaching. When you take on empowering others you will naturally be empowered too. The path of mastery is the path of a warrior. I don't mean in the common sense of one who engages in war, but in the passionate sense of one who seeks the greater self, for himself and for others, thus for the advancement of life itself. It's not the right way to be. It's only one possible way to be. It's the source of the fullest experience of being alive. I know. For the past 20 years, I've experienced this path, this way of the warrior. I call it the alternative to the decline of aging. It's a literal fountain of youth.

thought I was going to require resuscitation. But, I went back the next day and the next and I kept running day after day. Each day I increased the time and distance a little bit. Each day the quality of my running improved along with the level of my conditioning and recovery. Within three months, I could run several miles at each outing. I was 26 years old at the time. I became fairly fanatical about it. I was feeling better and better and better and better about myself. Fanaticism felt good! I kept running and running and running and adding more and more miles until I got to a point where I was running 8 to 10 miles every day except Sunday. Needless to say my weight started dropping over this time and within a year's time I got down to about 140 pounds. I lost 35 pounds during that year. I watched what I ate and I quit drinking soft drinks or eating or drinking any form of sugar, including fruit juices which some people actually think are healthy. Fruit juice is just loaded with sugar and we now know sugar makes us fat. My daily sustenance changed to include mostly fruits and vegetables with some occasional meat. In those days they didn't have all the exercise drinks we have now, so what I usually drank was water — pure water. I do that to this day. I love water and so does my body. Pure water is the drink of choice for the optimum functioning of your body. If we had gotten a manual with this human body, the instructions would have informed us pure water was essential to optimum performance and longevity of this miraculous biological machinery.

I was finally able to get back down to my ideal body weight. A few years later I began running marathon races, which are 26.2 miles long. In my training I became quite fanatical and got to a point where I was running 100 miles a week Kind of crazy, huh? My weight got down to a rather unhealthy 128 pounds. That is way below my ideal weight and I can tell you being underweight is not healthy either. I remember my mother-in-law visiting and asking my wife if I had cancer or something because I looked so bad. Then I learned about balance and even more science about our human machinery. Healthy fanaticism produces a thriving body. Unhealthy fanaticism stresses the body and will eventually cause a breakdown, or unhealthy imbalance, mentally and physically. I was in the best cardiovascular shape of my life but I was not healthy. I was pretty much addicted to running. It got

NOTES

to a point where it was detrimental to my health. It was a worthwhile experience because I now know the difference. I began to recognize I was in trouble when I began falling asleep in every situation you can imagine and I was always tired. My wife knew I was in trouble when I would fall asleep in the middle of a conversation with her. I adjusted to a more balanced and scientific regimen and again began to thrive mentally and physically.

I can save you the months of the wrong track in that experiment. What we highly recommend is you exercise daily and it's fine to be fanatical about finding time every day for one to two hours at the most. Work up to at least an hour of exercise every day and if you want a day of rest, take off Sundays. Remember to choose a form of exercise that is aerobic in nature, not anaerobic. You are not trying to become a world-class athlete — yet. Youthing is a journey and at various times in my life I have gotten out of shape and weighed over 165 pounds. That is what happens when you stop doing what works. Sorry, we humans are high-maintenance machines. I just again engaged in applying all the things we are teaching you here in the *Secrets of the Fountain of Youth*. I would faithfully use all the products we've mentioned in this book. I again engaged in an exercise regimen and just like magic I got back down to 135 pounds and in the best shape of my life. Find out what products worked for us. We'll send monthly updates on new things we discover. We'll spend the money and do the work and the experimenting. You reap the results and the rewards. We'll keep you posted about our stories and breakthroughs. We'll even provide an open forum for you to share your breakthroughs and discoveries with the rest of us. We'll be a community of warriors for life. We'll be friends. We really do want to hear from you and know how this is working for you. As new science and technologies emerge and we test-pilot them, we will reveal the results on the website. Our passionately committed vigil is to constantly and continuously search the world as new scientific discoveries are revealed to help us in our journey. Move it, move it, move it!

One of the fundamental elements of exercise is rest. I know that sounds funny, but rest is a key factor in avoiding injury when you exercise regularly. I call it R&R or rest and relaxation. There are days when you will push yourself to the max. You are under pressure at

NOTES

work, you have deadlines to meet, your home life may not be going smoothly, everything in your life could be amiss, and you need to know when to back off. If you add too much or too difficult a workout to a particularly stressful time in your life, and push yourself to new levels in your exercise, you may reach a breaking point. You must recognize your personal limits. It is true exercise may be the best thing for you, but include moderation and flexibility in your training. There are times when I take off and decide not to exercise.

Right now, as I am writing this book, I have taken off for the last six weeks and my body weight has gone back up to 160 to 165 pounds. I'm trying some other nutritional things I think are phenomenal and we'll give you the results on those as soon as the verdict is in. Life is constantly a test and how our bodies feel and function is a way to measure what works and what doesn't. What I found out over 30 plus years of running is the ideal length of time to do aerobic exercise, not just running, but any form of exercise is about one-plus hours a day. That little plus means that even an hour and a half on some days is good. When I keep my exercise to a maximum of an hour and a half, everything seems to work right. My body weight balances out as long as I watch what I eat. Everything just tunes and tones right up. I also just mentioned that I believe in R&R. After I take off for a month or so and don't exercise, I have to start out slowly and work back to the point where I can exercise for an hour or more. Then I painstakingly jump back in and force myself to start running all over again. I think this is an excellent process and I believe in it. I believe in the process of cycles in life. There are times in your life when it's okay to not exercise every single day. I know people who have 20-year records where they have not missed a single day of running, biking, swimming or whatever they do and I don't know if that is healthy or not. I believe moderation is really the key. I have taken off almost every single winter from any form of exercise. I've taken off at least a month and sometimes two or three months and let my body rest from all the pounding and other stress created by rigorous exercise.

Since exercise is one of the great keys to the secrets of the fountain of youth, let us tell you some of the benefits that come from it. We will also mention why it is one of the keys of life — here are a few of the facts. First of all it's a natural antidepressant. Any form of aerobic

CHAPTER 3

CREATING SPACE

Freedom Comes from Your Surroundings

What do you think I mean when I say, creating space? Fields of energy exist in physical space. During the last century, scientists found that space is a field of endless energy. Space is not vacant and void. Have you ever entered a room and sensed heaviness, discomfort, or even fear? What I'm speaking of here is not commonly thought of as physical in nature. You can't see discomfort or touch fear. They are indeed physical, though. You can feel discomfort, fear, or heaviness only because they have a vibration. You perceive these minute vibrations at varying levels of consciousness as well as the grossly perceptible ones. Don't we perceive upset even if someone is putting up a good pretense that he is fine? Most of us have never tuned in exercise is an antidepressant. If I were a doctor and somebody came to me with depression I would prescribe exercise. And I know it would work. I don't know of any case where it hasn't. Exercise is a safe and an effective antidepressant. It works. That's more than I can say for drugs. When you physically exhaust yourself running, biking, swimming, or doing aerobics, ballroom dance, or martial arts, your endorphins kick in and your depression, even clinical depression, can completely go away. When you continue to exercise daily it stays away. Have you ever listened to an ad for a drug on TV? Can you believe the possible side effects they are now required to tell you about? Exercise is a safe and natural alternative to dangerous drugs.

I used to think aerobics was for girls but my daughter and my wife got this step aerobic video and I tried it. I was running daily at the time and I wasn't even able to make it through the whole workout. Not only did I not have the coordination (I felt like an idiot) to do all the little moves, but I couldn't physically get through the workout. I now have a great new appreciation for aerobic exercise. I think step aerobics and all the other different forms of aerobics are a great way to get your endorphins to kick in.

We'll talk a little later about Pilates, Yoga, Tai Chi and all the other non conventional disciplines, but in the meantime getting started on any type of aerobic exercise is the important first step. If you are in an aerobics class or whatever form of aerobics exercise you are doing, remember optimally an hour a day will give you what you need in life. Getting back to the antidepressant benefit — it has been proven that any form of exercise helps the body start to produce endorphins. I've experienced endorphins and they are phenomenal. I have personally experienced that effect. Endorphins are an incredible tool. They are one of the secrets of the fountain of youth. As endorphins are produced in the brain after a certain level of exercise they actually help relieve pain in the body. They can give you a sense of euphoria you have only heard about. I've had what they call runner's high a few times in my life. Runner's high is what you experience when your endorphins are flowing and you feel like you can just run forever.

When I was visiting my family in San Francisco and attending a conference, I decided to stay in a hotel in downtown San Francisco.

with enough conscious and prolonged focus to be aware of the vibrations of the space around us. Most people are aware only of things, not the space in which things exist. I learned that the space around me is of primary importance. A human being and the space of the physical environment in which that human being exists are inseparably connected. Our physical environment reflects and impacts us. The physical environment can actually be an "existence system" for a past-based illusion for who we think ourselves to be. When we become aware of our physical environment and consciously create it to be aligned with a self-determined future, we can literally generate a new space of personal power. Our physical space then "pulls us" into that consciously created future and is the existence system to maintain it in reality. We can design an environment that allows for effortless expansion of personal power. I define personal power as the speed and effectiveness with which we bring our dreams into reality. Often we tolerate and adapt to the environment rather than take responsibility for it. If we take responsibility for it, we can consciously create it. When we create it, the space becomes alive with our energy and intention. Consider the possibility that your physical environment not only relates to, but provides access to a new freedom to be, clarity of thought, miraculous results, and powerful coordinated action. Buckminster Fuller said, "You cannot change people, but you can change the environment and people will change." Physical space is a mirror that reflects you and a force that impacts you. If you lift your environment to a new level, you will rise to that level.

What do you think it means to close down space or open up space? When we are experiencing harmony and joy in a space, it's easy to recognize as someone dumps anger and upset into that space, that space of harmony

Of course I ran the "killer hills" of San Francisco every day. I was running straight up these steep hills and then back down. I ran all the way to the Golden Gate Bridge and got hit with runner's high. I was running five-minute miles one after the other with little to no effort; I was just clicking them off. I don't know if you know how fast that is but doing one five-minute mile would out of reach for most people. I was jumping up in the air and doing 360s. I think most people thought I was on drugs. I felt like I could run forever. I ran about 15 miles that day and never felt sore at all. I ran as hard and as fast as I've ever run. I've had a similar experience two or three times but that level of endorphins is a rare experience. It's such a feeling of euphoria and superhuman energy it must be better than being on drugs. There is another type of high I've experienced many, many times during my daily runs. I experience these little bursts of energy where I feel just great. I know it's been instrumental in helping me keep my mental clarity and emotional balance. It has most certainly helped maintain that balance through all the little tribulations and even the really difficult trials of my life. The endorphins produced by exercise and the resulting experience of healthy mental and emotional functions are some of the most important reasons to exercise daily. Isn't it interesting as Americans have gotten more and more sedentary the incidence of mental illness in our society has also increased dramatically!

We have established one of the secrets of the fountain of youth is aerobic exercise. It is not only a deterrent to aging; it reverses the effects of aging. The reason is very simple. Your body produces something called human growth hormone (HGH) or growth hormone (GH). It has been scientifically proven that those who exercise regularly maintain a higher level of human growth hormone. Maintaining higher levels of human growth hormone also allows you to maintain your youthful appearance and energy level. I've been running for 35 years. I am 57 years old and many people guess me to be 35. This is not an accident of good genetics. It's the result of a lot of years of running and a lot of committed research to maintain my youthful appearance. A few years ago I was sitting in a room with a man and he kept calling me young man. It turned out he was 48 and I was 53 at the time. It is amazing to me exercise doesn't only slow down the aging process it actually reverses closes down. It's equally easy to understand that bringing filth into a physically beautiful space would also have a damaging or closing-down effect on that space. I dare say all of us have had experiences like these. We knew the space was uncomfortable, and yet we didn't focus intently enough to identify that the space had been diminished or closed down.

What is the impact of space that has been closed down? Consider that when the space is closed down it actually limits what can happen in that space. It can limit the feelings or thoughts you could have in that space. Perhaps we limit our experience of the impact of all that is really present in a space because we must deflect the literal impact of the bombardment of all that is there. It is difficult to be wholly present to all that is there in the experience of the space of our own homes, much less the space of the whole world. That we are unaware almost all the time of being impacted by everything in the space around us does not alter the fact that we are being impacted. You don't have to believe in gravity for it to have its literal impact on you if you step off a cliff. Another deterrent to being fully present to space is that in order to be fully present you also must then take responsibility for that space. That can be totally overwhelming. Can you imagine what it was like to take responsibility for cleaning up the space of Ground Zero in New York after 9/11? Responsibility for our space, although seemingly overwhelming, is also the access to being fully present to life and having the power to generate a great life.

Here's how physical space relates to the fountain of youth. If a human being is fully present to life and fully owning the space in which life occurs, it naturally follows that that human being will be more fully alive. When we become aware of the physical space in which we live and it. The youthing process must incorporate some form of exercise. If you just start exercising you have started the youthing process.

There has been a lot of research on growth hormone. Growth hormone is also referred to as human growth hormone or HGH. At the ripe old age of 23 the human body begins to decline in its production of HGH. Diminished HGH is responsible for the accompanying myriad of symptoms of the decline of aging. The best way to increase production of growth hormone in your body, of course, is for the body to produce it naturally and exercise is one of the best ways to do that. It is the access to staying young even in your old age. If you take up running or any of the other aerobic exercise regimens we've suggested, it can actually help reverse the aging process. Our term for that process, as you already know, is youthing. Exercise is one of the secrets of the fountain of youth; if you exercise you will grow younger. Look at anybody you know who has exercised and notice if you don't think they look younger than their years. I guarantee you that you will find it so across the board. Even if they haven't exercised for 20 years, as soon as they begin again, it brings back an appearance and an energy that is best described as the vibrancy of youth. Truly, truly, truly this is one of the secrets of the fountain of youth. We strongly suggest, encourage, and implore you to begin a consistent exercise program today. We encourage you to begin an exercise program and have a body you can love and appreciate all the days of your life. Even if you have major physical limitations, you can improve to the point at which you can exercise again. We suggest you believe in exercise as we believe in it. It is the key to having a vital life, and to youthing; to having a life you love. Any form of aerobic exercise will do. Find one you love and it will put you on the road to the fountain of youth.

Finding and Using a Trainer

One of the surefire ways to get into aerobic exercise and to keep on exercising consistently for years is to find a qualified trainer. A trainer is a personal assistant who can help you start training and keep training in the future. Trainers not only know about aerobic exercise, they know about anaerobic training to supplement your aerobic exercise. Most trainers are experts on nutrition and avoiding injuries. We believe in are responsible for generating a space of order, a space of beauty, and a space of self-expression, the space becomes very "quiet." It does not pull our energy down or impact us with a feeling of too much to do and overwhelm. The space then nurtures us and becomes a space of completion. Completion creates more space. We then have a sense of control over our lives and have the space to expand. When we lift a space with our diligence and creation, it then lifts us to a new level of knowing and expressing ourselves.

My husband and I bought an incredibly beautiful home in the Wasatch Mountains of Sundance, Utah. If you know anything about the prices of homes in Sundance, Utah, and even if you knew that my husband is a successful consultant, you would also know that buying this home was an unbelievable stretch for us. Going for the glorious in life usually is. We even bought this home before selling our other house in the valley. We had also bought a piece of property where we thought we wanted to build a new house and even went so far as to have an award-winning architect do plans for this house. The lot had a beautiful view of the city below and the Wasatch Mountains behind and was conveniently located halfway between Orem, Utah and Salt Lake City. Ron's travel time to the airport would be cut in half. It made perfect sense to build there. Even with a beautiful view, it would still be just a suburban house in a suburban neighborhood with a great view. It would be a beautiful house, but it wasn't our dream. The problem was we loved the wilderness of Sundance. Whenever we went for a hike it was up in Sundance. The experience of being in Sundance is one of being surrounded by the rare beauty of pristine forests, waterfalls, glaciers, and some of the most interesting people on the planet, while only 20 minutes from town, with all its accompanying benefits. With Ron still traveling so much, and the heavy snow in trainers. I always say, "I would never make myself do what a trainer asks me to do." For a more detailed explanation and a list of qualified trainers worldwide, go to www.secretsofthefountainofyouth.com and click on "Trainers."

The journey or road to youthing is not as difficult as you might have imagined. Exercise is one of the accesses to youthing. To do nothing is to inherit the decline of aging by default. Exercise is life! We have personally experienced the benefits and results of exercise. We know you can do it too. We recommend you start now and persevere until you can maintain at least an hour to an hour and a half a day. We know a personal trainer can support you in your commitment and teach you what you need to know to be most effective in the process.

the Utah winters, it didn't make good sense to move to Sundance. We kept creating the practical solution, but for one reason or another we just kept not starting to build the new house. Finally we told the truth about what we really wanted and took action to make it happen. This is underlined because it's one of the secrets of the fountain of youth. It's the space of true integrity. It's ultimately being true to yourself, true to your most desired self-expression, true to your heart's desire. Just sit quietly or engage in a committed conversation with your partner or family; do that long enough and your truth will emerge. If that truth happens to be consistent with reasonable and convenient, it's only by accident.

Sometimes we really want a lot of things and we don't get them. It's hard to get that we don't get it only because we are not willing to give up the view of ourselves as not being able to have "that" and we are not willing to do the work to get "that." We kept pushing through all the seemingly impossible obstacles to our having that house. If you want training in accomplishing the impossible you should have my husband for a coach. Even though he is a consultant and does coach people all over the world to extraordinary accomplishment, getting this house seemed so impossible, I questioned our sanity in taking it on. So did Ron. He was truly amazing though, in putting that whole deal together. It was brilliant.

After we miraculously managed to get the house, Ron went into a bit of a tailspin. It was something like, "What have I done? I must be nuts! I have three property payments and the same amount of money coming in. I'm going to bankrupt us. What was I thinking?" Our old house didn't sell and didn't sell. Ron worked harder and somehow kept covering everything, but his blood pressure went up for the first time in his life. He's been an athlete all his life, you

CHAPTER 7

FOOD AND NUTRITION

Modern Day Miracles

Let's talk nutrition and supplementation. What nutrients and supplements are essential to youthing? Every single person in the world must have heard the phrase, "You are what you eat." As little as 150 years ago, people mostly grew all their own food. People worked hard. It was forced exercise. They tilled, planted, and tended their gardens and fields, growing their fruits and vegetables organically and naturally. They picked them fresh, ripe and full of nutrients. Even their methods of food preservation were natural, dehydrating food in the sun, preserving most of the nutrients. Their lifestyles gave them a far better source of nutrition than we have today. If you'll pardon the pun, when we look at the fruits of our labors, we are a society that is one-third overweight, and one-third obese. What happened?

In our modern world the pace is fast and we spend our days and nights working, worrying about financial and family problems, and stressing over many things. How do we maintain health and energy in the face of an incredibly stressful lifestyle? Let's first look at the food we eat.

remember. He went to the doctor, who didn't find anything physically wrong with him. The doctor asked him if he was under any pressure lately. That was an understatement. We were still living in the house in Orem and I was working on the Sundance house to see if I could get it ready to rent out during the ski season. I moved about three fourths of our furniture up there and completely furnished the kitchen, the bedrooms and baths. Fortunately, we had seven bedrooms in Orem and only three and a loft in Sundance, even though the Sundance home was bigger. At least we still had beds to sleep in back in the valley. The house in Orem still didn't sell. Finally it was our son, Adam, who said, "Why don't we just move to Sundance?" I looked at him and said, "You know, I think you are right." Every day I loaded up the SUV and settled that load of our things in the Sundance house. Within a month we totally moved. One day Ron called from the airport and I said, "You live in Sundance. That is where your clothes are. Come home to Sundance." Even though our home in Orem was in excellent shape there were some things I never liked about the house and never fully tuned in enough to my dislike to take responsibility to do anything about them. I sensed we needed to upgrade the house to compete with the new homes for sale in the area, even though our house was only 11 years old. This wasn't exactly a good time to be spending another $5,000, but that is exactly what I told Ron we needed to do. I knew some things about space and saw the results of raising the level of space and having "stuckness" dissolve, allowing for miracles to show up. He trusted my intuition. He's such a smart man. In the middle of moving to Sundance and contracting to have an outbuilding completed for Ron's office, having the kitchen floor refinished (never refinish the kitchen floor after you have spotlessly cleaned the entire house), and shopping for Ron's new office furniture,

One of the key factors contributing to our obesity, our being overweight, our rampant type II diabetes and other diseases, is directly related to the food we eat. What we eat and how much we eat is also the key to turning all that around. It's an essential key to youthing. There is no question about it. Remember one of our fundamental philosophies is moderation in all things, not about being fanatical. But it's also about reality when it comes to food. We need to get back to the basics in the domain of the food we eat if we want health and vitality all of our time here on earth.

Let's take a look at the typical schedule and diet of a city dweller. Typically we are jarred awake by the alarm clock, jump out of bed, mechanically rush through our morning routine of hygiene and dressing, and head off to work, almost always running a little behind schedule. We immediately get irritated at the delays of traffic and the idiotic and rude drivers out there. We need to hurry yet stay just under the speed at which we won't get a ticket. About 50 percent of us don't even take time for breakfast. Those who do, usually scarf something down that isn't healthy. Often it's something we pop in the toaster or microwave that's full of sugar and white flour. Sometimes we grab a quick bowl of cereal. You eat some highly processed cereal that no longer even resembles a grain. It's dyed to a supposedly attractive color with artificial food coloring and coated with sugar. You throw some milk on it, scarf it down and rush out the door. That is our traditional American breakfast. Lunch isn't much better. Fast foods are the rule for our fast lifestyle. Fast food is again highly processed and almost always full of fat and sugar. It has almost no nutritive value. It just satisfies our cravings for taste and fills our stomachs.

We need to return to eating food that is as close to the natural state in which it grows as we can. Breakfast and lunch are the two most important meals and most of us don't realize how important they are. Most of us think dinner is the main meal of the day. Sometimes after paying almost no attention to the quality and quantity of nutrition we have gotten all day, we come home and gorge ourselves on steak or some other form of meat, very few vegetables, and nothing fresh, unless you consider produce from the store fresh, which I don't. Then you indulge in some sugary dessert. You've had a hard day and need to

I bought new stainless steel appliances for the Orem house and contracted to have two bathroom floors and the laundry room redone in stone tiles. I bought new light fixtures for the bathrooms and had the family room re-carpeted. I even replaced the front door handle, the doorbell, and the front door kick plate. I spent about eight hours replacing all the old cabinet door pulls with brushed stainless steel ones to match the new appliances. A landscaper owed me money for his children's Taekwondo classes and I called in my marker. He did an incredible spring cleanup in the yard…in between late snowstorms. I bought all the materials and found workmen to do the projects in both places, all in the middle of the heaviest snowfall in Utah in 10 years. More than once I had to help a workman dig his car out of the snow when he was trying to get up to the Sundance house without four-wheel drive or chains. It was too hard to be funny, even looking back on it. I was still moving loads of household goods up the canyon every day. I felt like a pack mule that moonlighted to plow the north 40 at night, but the Orem house sold the day I got everything complete. Miracles are not without work is the moral to this story. The other lesson here is to "listen" to the space. Just pay attention. Walk through the space and listen. Let yourself imagine. Images of what needs to be done to lift the space will appear-if you are looking, and willing to tell the truth and act consistently with that truth, not just do what is practical and reasonable. Miracles don't live in the domain of practicality and what's easiest. You'll find them appearing in the domain of truth, no matter what the cost.

I call myself a "dreamer-doer." If I envision something, I am willing to do the work to bring it into reality, no matter how long it takes, no matter how hard, and no matter how much it costs. I've noticed that income and time expand with desire and demand. The impossible is only relax. We sit down in front of the TV and have a glass of wine or beer. Very seldom do you drink water. After the wine makes us sleepy, we head off to bed to start the process all over again the next day. Kind of sad isn't it? We not only need to stop the endless treadmill long enough to consider our nutrition, we need to consider our lifestyle. We enjoy a society of more freedom than almost anywhere on earth, yet many of us don't take time to consider we even have choices, much less take responsibility for making them.

We want you to go to the website www.secretsofthefountainofyouth.com and click on "Nutrition Log." Download the chart and use this chart to journal what you eat and drink for one week. Basically all you do is write down everything you eat and drink every day for a week. Don't let this chart scare you into eating something healthy; that is not the point of this exercise. The point is to become aware of what you currently eat and drink. What you do is write down what you eat and drink every day for a typical week, Monday through Sunday. Every time a liquid or a morsel passes through your lips, you record it. You write down *everything* you eat or drink. Record your snacks, candy bars, or a single bite of chocolate. Record that bowl of ice cream you had late last night, or the coffee, the milk on your cereal, the beer or whatever else. Record it all. Write it down on the chart. This should give us a pretty good representation of your diet. If you truly are what you eat, this should be very revealing. To become aware of what you eat will give you power. You can't create until you know where you are. Then we can analyze your diet and help you choose what and how much you are going to eat in order to create the body and level of vitality you desire.

After you are done with this week long chart, you can do the analysis or you can finish reading the entire book and do the analysis at the end. We are going to analyze what you eat. We are going to do it in a very broad way that includes all possible styles of eating. You will be able to analyze your diet from a very broad point of view. You will recognize your style of eating and see what kind of changes you need to make in order to continue your youthing journey. Your weight and your youthing really are directly related to what and how much you eat. What do you eat and how much do you eat?

After you have recorded a whole week and written down everything you ate and drank we can look at each day and categorize accomplished through impossible work being done. You must live in the creation, not in your thoughts about how hard it all is. I would occasionally indulge in a little self-backslapping about how hard it all was and how magnificent I was in the face of it. I suppose you could interpret that I made the house prettier and the pretty house, of course, sold. That interpretation doesn't have as much power in it. I know I listened to what the space needed to be complete. I sold our house in California the same way. I turned an outbuilding into a greenhouse. I thought it was a good use of the space. I was moving; I didn't need a greenhouse. I was inspired to build the greenhouse. The last nail was literally being driven in when someone stopped to see the house. She was an avid gardener, fell in love with the greenhouse and made an offer on the spot.

In my kitchen in Sundance two walls and a third of the ceiling are glass, all looking onto the most magnificent view of a wilderness forest and waterfalls tumbling down the face of an 11,500 foot snowcapped mountain (I live in a postcard!). The kitchen cabinets are all glass, which makes everything in them visible. The "feeling" of the room was one of clarity and a clear view. Do you remember me saying that I had to furnish the kitchen in hopes of renting it out for the ski season? Well, I was still using our plates and glasses, and so on in the Orem house, so I had to buy some new ones for Sundance. I looked at many dish sets that looked rustic like Sundance and nothing really inspired me. Then I saw some inexpensive plain glass ones and I felt they were the ones. When I started putting the kitchen together I couldn't make myself put anything in those cabinets that wasn't clear glass. That wasn't hard, because I always liked cut glass and lead crystal and I inherited a few nice pieces. The cabinets are lighted within and below. With only clear glass in them they radiate brilliant light. People remark the foods you have eaten. What we want you to identify first are the broad categories. We want you to reorganize all the food and drink you ingested during this week into these broad categories and record your findings onto another chart. Go to the website and find the Food Category Chart.

As you are able to divide your foods into these several different categories you will see a pattern emerge. Identify your individual pattern. Once you see your habitual pattern of eating you will have a scientific basis to start making the changes that give you optimum health and extraordinary performance. You can take control of this part of your life. Your overall dietary patterns will be clearly visually demonstrated on this chart. Be sure to measure and include the quantity of what you eat and drink. Just consider approximate ounces and servings to estimate quantity. Quantity is as important as quality. It's one thing to eat a Big Mac and some fries and a milkshake, and then there are those people who eat two or three Big Macs, two or three large fries and two or three milkshakes — no kidding! In regard to what you eat, we want you to ask yourself several questions.

Please look at this chart as you read the questions. The first thing to analyze and the very first question you ask yourself is, "Is this healthy and does it contribute to my youthing or aging?" That is the very first question. Take a look and decide which foods and drinks are promoting your youthing and which are taking away. I assure you that you do know which is which. If the answer is negative, circle those items. You will then be clear about what to eliminate if you are choosing aliveness. Begin by eliminating some of the things you know to be the most detrimental to your wellbeing. Then we'll look at what foods to incorporate into your new style of nutrition that will reverse the normal decline of health and performance associated with aging. We want to produce variety and balance in what you eat. You notice I did not say, "Eliminate all those things." I said the word "balance." Too many diets are fanatical and take the fun out of life. Our objective is to have extraordinary health and performance without taking the fun away. One of the keys and secrets to the fountain of youth is striking a balance and learning to eat healthy food in quantities that promote health. The process of getting clear how the kitchen seems like such a clear space and what an amazing feel there is to the room. Don't think that these are some kind of decorating tips. I have no background in decor. I call this "reading" the space and taking appropriate action to that. It's not a talent. Anyone who tunes in will see it's so. When we are true to that awareness, the space takes on a magical quality that uplifts and magnifies all who enter. My Sundance environment is changing me. My life has been so frenetic for so long, it wears most people out to be around me. I am so much more focused and peaceful now. People feel that now when they are around me. I don't watch as much TV. I used to be exhausted at night after running around all day, that I would want to vegetate in front of the TV for four hours. Now I write, read, and listen to good music. My family stays home more and we have more friends over. I am "called" to cook in my incredibly beautiful kitchen. I'm inspired to make a beautiful table. My cooking has become more creative, rather than just getting us fed. With the clean air, spring water, and morning mountain hikes, I can't help but think this environment will have a huge impact on our health, well-being, and longevity.

Our environment has made us all feel like life is even more special. In contrast, if someone brings conflict, bad mood, or negativity of any kind into this space, I have zero tolerance and will put in a correction right away. I'm making different choices. My husband and I have always had our own separate interests and work. Now I am working with him in managing the family finances and being his partner in developing a seminar for people over 40 years old. It's called "Winning the Second Half Summit." My husband always wanted to do more with me and especially be partners in a project like this. It's his dream come true. I'm finding it's mine too. We are even experiencing being about your diet and scientifically categorizing what you eat gives you a clear, logical space to choose to eat those things are going to make you healthy, feel better every day, increase your energy level and thus put you solidly on the road to youthing. It should be obvious to you that a great deal of the things we eat and drink are not good for us and it's time to understand that quality of life depends on choices. If we really want our youth and vitality back, we are going to have to make some of those choices that will directly influence the very quality of our lives. Vitality, health, and strength are daily choices. These many daily decisions of what to eat and drink and in what quantities are not just things to think about while reading this book, but are actually choices to make many times a day for the rest of your life. If you are choosing aliveness, it's time to become aware of just what you are eating and start eliminating those things from your diet that aren't good for you. You are not just eliminating those things from your diet but are eliminating the results of eating those things from your life. Remember, "Garbage in equals garbage out."

Taking responsibility for the foods you choose to eat daily is a key element to youthing. There are many good books on nutrition. Educate yourself on human nutrition. Would you buy a car and not read the manual? Would you guess about how to maintain that car and about what fuel or oil to put in it? How long would that car last without a breakdown? Without consulting an expert, you really know every time you consider what to eat or drink you have the power to decide whether it's good for you or not. Even without reading a book, you really do know that live food produces aliveness and dead food produces death. The problem lies in not clearly confronting what we are choosing and the profound impact of our choices. The most important distinction about food is simply to choose what you eat, so choose wisely. You can always refer to our website www.secretsofthefountainofyouth.com.

I again remind you to study the glycemic indexing link. It is a great tool. Educate yourself about the glycemic index of foods. It will help you to make good choices — choices for optimum health and aliveness. We encourage you to engage in making the choices and doing the work to always be youthing rather than aging. Refer to chapter 5 often.

more and more in love. In making this move to a whole new level of life and calling forth having what we have always wanted, we have discovered what a good team we are. I am very good at creating space with my organizing and inspired design. Ron is very good at generating the money to do it. He creates space for me and I create space for him and so on and so on. It's definitely an empowering upward spiral. I call it a power-feeding relationship.

Maybe it's not the time for you to live in a mountain retreat, but even if you live in a flat in New York City, you can create it as a self-expression. That means you can have around you the things in life that uplift you and validate who you are. What I mean by that is to not tolerate hanging onto to junk just because you don't have the money to have what you really want. I prefer to have an empty space rather than fill it with "make do." If that is all you can do now, get one really special treasure at a time. "Listen" to the space and wait for inspiration. You will know when you have the perfect fit. When you achieve that perfect expression, the space will call you to the next level of your life. You will find new opportunities and possibilities showing up.

Creating space has an enormous impact on reality. It's a commonly known fact that environment has a significant impact on productivity, mood, and satisfaction. My husband is a business consultant. One of his first moves in consulting with a company is to analyze the physical space. It's sometimes amazing to me how businesses can have any measure of success when they are not organized for the flow of the business. People are heroes to produce what they do in sometimes totally unworkable environments. Sometimes we walk through the same piles of unconscious disorder and we no longer see them. They impact us even if we don't see them. We inherit many environments and

You must determine what you need to eat more of or less of. You need to identify food or drink to which you know you are addicted. You know these addictions cannot be good for your health unless it's an addiction to exercise. Many people are addicted to different foods and drinks. Obviously, if you are overweight you are addicted to some foods that are not good for you. If you are obese you are dangerously addicted to certain types of food and probably sugary drinks. If you choose aliveness and youthing, identify your addictions and exercise your power to eliminate them from your life. Sometimes just being aware of them will do the job. When Mary Louise saw a dark-field microscopic picture of what drinking a diet soda did to her red blood cells, she went cold turkey that day. After years of her husband encouraging her to not drink diet sodas, seeing clearly the damaging effect to her body "cured" her. Ignorance is not bliss! Ignorance is death! It's not a fun death either. It's an existence in which you are not happy with what you see when you look in the mirror every day. You are miserable about how you feel and what you can do. It's a slow, tortured death. Choose aliveness. Choose youthing.

Sugar

The brain has the power to generate your getting everything you want. The experience of having control is an issue of choice, not possibility. You have control of your future and your life if you choose to own that power. Controlling what you eat is one of the most important keys to your health and your youthing journey. Let's talk about the choice to consume sugar or not. Here's a bold premise — anything with sugar in it is bad for you. What's funny about that is everybody on the planet already knows sugar is bad for you. It is no secret. Sugar is poison. High glycemic or low glycemic, sugar is the enemy. Every scientific study we have examined, our own experience, and our observations of others lead us to believe sugar is poison and if you eat enough of it, it will kill you. The first thing to look at is how much sugar you eat.

Sugar is an enemy to health. We state that unequivocally. Sugar is the enemy. Sugar makes you fat. Sugar wrecks havoc on your biological and biochemical systems. I know this from research and from my own experience. I like sugar as much as the next person. It tastes good. I crave it. Yet, every single time I eat sugar before I go to sleep, every single can feel hopeless to change them. That is why you need a community working together to take on this kind of a transformation. On the other hand, space can be very orderly and even well designed and still not be inspiring. Sometimes it's only some small, yet personally meaningful expressions that will alter the entire space. After Ron experienced the miraculous expansion in his own business when I totally cleaned, organized and designed (C.O.D.) his office, he began taking me into his clients' companies to transform their physical environments. It allows him to make much more of a difference with his consulting. He gets excited when he sees piles of paper and things stuffed anywhere. He knows when we shift the space, big results will follow and he'll be a hero. We now go in, and along with all of the employees, do a "SpaceWorks Intensive" a workshop delivered over two or three days. The atmosphere is akin to an Amish barn raising. The additional bonus is building a work community of coordinated action and real team. It seems like people are blown away by realizing how much they tolerated before, thinking it couldn't be any different. When they begin to focus on the environment, looking for workability and how to create uplifting beauty, they see right away what to do. What one doesn't see, another one will. One will have half of the idea and the other will see where to go from there. When people participate in creating their environment to be exactly what their eyes love to see and create the space to support the flow of their work, they are incredibly more satisfied in their work as well as being much more effective and productive. They seem to "own" the space after that. Not only does the space become personally theirs, but so does the business. The company becomes one of responsible partners. One company's bottom line improved 1,000 percent in one year after their SpaceWorks Intensive.

time — not just once or twice or occasionally, but every single time (did I say that enough?) — every single time I eat sugar before I go to sleep, when I wake up the next day, not only do I not sleep well, but my brain is groggy, my muscles are sore and not responding fully, I feel old and sluggish, I have no energy, and I have gained weight. Sugar is the enemy. Everybody should now know that. We go on record stating that sugar is the enemy of health, plain and simple. So sad, me too, too bad! If there is one thing to get from reading this book that would make the biggest difference to the quality of your health and performance it would be to eliminate sugar from your diet. It is one of the most important choices you could make to give yourself extraordinary health and optimum performance. It is one of the fundamental secrets of the fountain of youth. Let's say it one more time and another way. If you want optimum health and wellbeing — don't eat sugar. Did you get it? Say it with me, "sugar is the enemy of health!" At least if you eat it you'll know you have chosen illness instead of wellness this time. Then you can correct and get back on track. At least you won't be kidding yourself. It's really a bad joke! To have or recover your health you must eliminate sugar, both overt and covert, from your diet. Read labels! They hide sugar everywhere! They want you to be addicted to their product. I'm in trouble! I've squealed on them. I'll have to look over my shoulder for the rest of my life. You have no idea how much money is involved and I've not only given you your health back, I've given you your money back. I've taken your money back out of their greedy little hands! You see — that's what they want — MONEY. They don't care about your health — you'll have to do that. They don't care if they kill you. That's just collateral damage.

Remember most artificial sweeteners are worse than sugar itself. Don't be fooled by that ploy either. They figure if you wise up about sugar making you fat and sick, they can sell you artificial sweeteners. That's the code name for deadly chemicals. There is only one replacement for sugar that makes any sense, Xylitol. Xylitol is a sugar made from plants. It is very safe even for diabetics. Go to the website if you want more information on this technology.

Does that mean you can't ever have a cookie, a piece of pie on Thanksgiving, some ice cream, or even a piece of chocolate fudge now

In discovering a new power in life by being responsible for consciously creating our environments, I observed a flaw in our focus. We human beings have been viewing ourselves as "things" — and we are, but that is not all we are. We have mountains of data about this human biological machine, but what is the source of life and energy? Why do some people say they have no energy when there seems to be no biological deficiency or explanation? We are really spaces existing in spaces being impacted by other spaces. Who we are impacts space and we are impacted by space. We are profoundly connected to the spaces in which we exist and for the most part have been unaware of that. We live in the illusion of being disconnected from the space in which we exist, even the space of our own bodies, and just use the space. We haven't realized that in taking responsibility for space being workable, uplifting, and complete, restrictions on our personal power are dissolved. The experience seems to be one of being more in control of life. New passion, energy, and openings for action show up in space that is designed to work, nurture, and inspire. Human beings are most alive when the space in which they exist has been designed for that. This is not a difficult theory to test. Try it out. If you are doing this on your own, first take on a small space. Get a friend, spouse, or colleague to support you and co-create. There is more creative power in numbers. You will also have more courage to take on a really overwhelming space with support. With the help of my best friend, Susie, whom I cruelly put to work on her vacation, it took eight to ten hours a day for eight days to completely bring my husband's office to the level of which I speak. It was right before Christmas, and Ron was traveling, so I put a big red ribbon and bow on the door, announcing my gift to him. In the next and then? **No!** Aren't you happy now? We took it away and then gave back — a little. Remember, we also recommend moderation in all things. Just remember to have it so much more then than now. Have only small amounts and savor it longer. There are too many people addicted to sugar and we mean that in every sense of the word. There is nothing wrong with having an occasional dessert of something sweet — just don't overindulge. Don't let sugar control you — you control sugar. When you do it only on rare occasion it won't become an addiction. Just like alcohol, if you can't eat one piece of fudge without eating 20, then don't eat it at all. Your body is more easily addicted. Just as a person is addicted to cocaine, other drugs, smoking, alcohol, or coffee, many, many, many people are addicted to sugar. Do you remember what I said? Sugar is the enemy! Sugar and sugar substitutes should be eliminated from your diet as much as possible. If you never ate anything or drank anything with sugar in it again, you would be better off. You are definitely on the road to youthing if you can cut sugar out of your life. Many people don't want to hear that. We'll say it like it is. Sugar is an enemy of youthing.

When we talk about sugar we mean all forms of sugar. You can even eat too many fruits. Once I went on Nathan Pritiken's diet for about six months. I never felt better in my life, but it was one of the hardest things I've ever done. The diet does not allow you to eat any sugar or fat — zero! I never felt or looked better in my life, but it was not easy to do. What we are suggesting is you just get your sugar intake under control. If you are able to do that, you will feel the difference. If you are able to totally eliminate sugar and all its derivatives, you will absolutely feel the difference. I see a lot of people order diet drinks. That not only doesn't work, it does more damage than sugar. Diet drinks are just as bad for you, if not worse, than regular soft drinks. Sugar and all of its chemical alternatives are bad for your human biological machine. Enough said. I think you got it!

There are other addictions you should be aware of and should avoid like the plague. Smoking is an obvious one. It's not a food but you know what I'm talking about. Sooo many people are addicted to beer and other alcoholic drinks. We have an even worse opinion of alcoholic beverages than we do of sugar. Alcohol is sugar and it's not three months following that physical transformation, his results went off the charts! Don't take my word for it. Put it to the test.

The first thing to do is "be" in the room, looking. Have a notepad and jot things down as you recognize them. Ask yourself and your partner, how you feel when you see "that" or "this." You'll probably come up with answers like, "heavy," "overwhelmed," and "closed in." I remember my husband had a shelf over his desk and it must have had 40 books on it, most of which were books he had not yet read. When I asked him how he felt when he looked at them, he said, "It feels like I have so much to do." I put the books in a closed cabinet and made displays of some of his favorite things from his travels and then asked him how it felt. He said, "It reminds me of what a great life I have." Which one do you think will give more energy? Our environment is often a huge sucker of energy!

There are a few basic principles to follow regarding "clearing up" physical space. You want to first analyze the space and identify what doesn't or does work about the space. Brainstorm a little about what you would really like to have around you. Label a few boxes as Throw Away, Give Away, Repair, For Display, Archive, just to name a few. You will come up with more according to your need. Then clear everything out of the space. Next, you will clean. You will clean in a way you were never willing to clean before. That is why you have to take on small areas at a time. The level of detail can be unbearably tedious and you want to open your eyes and really see the detail. When spaces are dirty and cluttered, one of the ways in which we deal with them is to tune them out; we do not see them any more. They are still impacting you. You should include in your cleaning supplies carpet spot cleaner and Goo Gone for removing old gummed labels, tar, crayon, ink, and so forth.

good for you. The destructive nature of alcohol is so much more than being treated by the body as a simple sugar. Alcohol can destroy lives — not just yours, but every life you impact. That impact is with your car if you drive under the influence of alcohol. Many major, life-altering mistakes were made under the influence of alcohol. Mary Louise once talked to a minister in the Amish faith. He said they didn't smoke, drink alcohol, or take mind-altering drugs because they wanted to be good. He told her the reason Amish people don't drink alcohol, smoke or take any mind altering drugs is it makes the human spirit "coarse" and you can't hear God trying to teach, guide, or bless you.

Finally, many people are also addicted to caffeine. Some people have to have their coffee "fix" every single day. Some people say it helps you run better; it helps you exercise more. This is horse manure. Caffeine is addictive even in minute amounts and it is not worth whatever it does to your body. Caffeine has no place in the human body and the way it harms you far outweighs any perceived benefit. Does that mean you can't drink coffee? Yes, if you are addicted. But again the moderation thing is what we are talking about. Some people are truly, truly addicted to coffee. Check the Starbuck stock. People are spending millions on their coffee. Can you believe people will pay four bucks for a cup of coffee? Yet go to any of the coffee shops around the world and you'll see people standing in line every morning. It's an addiction and not a healthy one either. Caffeine is an addictive substance. If you are addicted to coffee or any other caffeine product, you already know it. It won't make you younger.

Some people think orange juice is good for you. Orange juice is processed and during that process they kill off everything that is good for you — all the enzymes and most of the vitamins, minerals, and phytonutrients. Then they put back artificial vitamins and minerals and tell you how good it is for you. Orange juice is a condensed sugar. Eat a fresh orange. Orange juice also causes a significant acid-base imbalance. Your body, by the way, is alkaline. As we mentioned in our discussed coral calcium, your goal is to achieve an alkaline balance in your body. It is the environment of health. An acid environment is one of illness. What you eat and drink greatly impacts your body's acid-base balance. When you eat a fresh fruit, like an orange for example, it has natural

Goo Gone is a great product. Don't buy similar products with another name. Many of them will damage finishes and paint, whereas Goo Gone won't. I had to learn that the hard way and like our intention for this whole book is to save you from the "hard way." The need will let you know what else to assemble. Buy a labeler and do printed labels for everything-no hand-scratched labels. You want to go for uniformity. Uniformity is quieter and gives a sense of order and control. You are going for creating a workable, uplifting space. Take everything out of the space first. With everything pulled out you will be able to thoughtfully and consciously create the space a new. After clearing out and cleaning up, plan and organize the space. You will probably have to buy organizers of many varieties. They don't cost much compared to the results I promise you will get. Stores like Hold Everything will have some really creative and workable organizing products. For plain bins, go to discount stores, where you will get the best price. Again go for quiet and uniform. I always go into a company the day before a workshop and see what we will need and have it ready to go the next day. Then we don't waste time waiting for supplies. Assemble everything before starting the job. The idea is to contain everything in a workable, non-distracting or heavy way. I also say that whatever is worth posting or displaying is worth framing. Remove all the Post-It notes from the walls or computer. The cluttered, busy feeling will disappear right away just doing that. Buy a small notebook to capture ideas, appointments, or tasks as you think of them. Ask yourself the question, "How will this space work best for the function for which I need it?" After planning and organizing the space for function and flow, put in your personal design. You can start with a "radiating point." A radiating point is some enzymes that convert that acidity back to alkalinity. When you drink processed orange juice it does not have those enzymes anymore. They were destroyed in the processing. Processed orange juice is 4.0 on the acidity scale, has a lot of sugar in it and is not good for you. Processed orange juice is one of the worst things in the world for you. Even fresh squeezed orange juice is not good for you because it is a condensed sugar. Eat an orange. Eat it as nature intended. That's a good rule of thumb across the board.

Alcohol and coffee, which are 4.0 on the acidity scale, are also bad for you because they cause an acidic condition in your body. It takes your body 24 hours to balance itself after only one cup of coffee. Your body is working so hard to keep you alive. Do it a favor and help it out with some good choices. You need to remember that anything that causes your body to get out of whack is not going to be good for your health. Orange juice, alcohol, and coffee are the three most acidic drinks on the planet. I'm sorry to be the bearer of the bad news that so many things are bad for you. Extraordinary health is actually more fun than indulgence.

All those things having been said you might ask what *is* good for you? Let's move on to the positive side of nutrition. What are the things are good for you? What are the things you should eat a lot of? Obviously everybody hates to hear it, but salad, raw fresh vegetables, fresh fruit, and any other wholesome herb or plant-based food is good for you. You can eat up to five fresh fruits a day. You would be extremely healthy if you could eat fresh fruit right off the tree. Some of the things go through the grocery store, such as apples, are okay. Other things grocery stores keep on the refrigerated shelves for weeks lose most of their nutrition. They really stretch the definition of fresh vegetables. Apples happen to be one type of fruit you can store for long periods, even months, at the proper temperature. Even then they maintain their living enzymes and the phytonutrients that are good for you. The only fatty vegetable there is, and it's not really a true vegetable, is an avocado and even an avocado in moderation isn't bad for you. It's actually one of the foods richest in omega 3 and omega 6 fatty acids. But everything else is fair game. You can chow down. Fibrous vegetables will never make you fat. You can't overeat on fibrous vegetables. Did you notice that word personal object that is precious to you. When you see this object you feel good, at peace, or even joyful. This object can actually tell you a lot about yourself. One of my radiating points was a beautiful necklace my husband had given me. You choose that certain precious personal item, and identify the qualities of that item and how you feel about it. I would say these qualities out loud to another person or write them down. If you look at those qualities independently, they will reveal to you what you value most in life and how you want to be. My necklace was made of diamonds and pearls. It was handmade and one of a kind, so it was special. I like to be special. My husband gave it to me and I especially like to feel I am special to him. The necklace is brilliant, beautiful, and unique. I would like to be brilliant, beautiful, and unique. They are of heirloom quality and therefore precious. I must be of great worth to have something so precious. Now this is all interpretation, yet that is really at the heart of why certain things are precious to us. Often we put precious things away in a drawer; kind of how we sometimes guard the most precious things about ourselves. That is where I put my necklace. When I got this concept, I went out and bought a beautiful glass box in which to display my necklace. Now every time I see it on my dresser I am uplifted. I once had a client who was part American Indian. She had some beautiful handmade Indian pieces of art she inherited from her grandmother. They were stored in a closet, being saved for I don't know what. We made several displays of her Indian heritage and it brought her to tears. We created shrines to some of the most precious aspects of her life. Every day it reminds her of her heritage and her grandmother, who was so special to her. When people are surrounded by who they really are, they are more alive. More alive is definitely a secret of the fountain of youth.

fibrous? Fibrous is distinct from starchy vegetables. Potatoes and corn are examples of starchy vegetables. You can over eat starchy vegetables. Too many starchy vegetables can make you fat. They still have great nutrition but, too much will make you fat. Vegetables are great. Salads and raw vegetables are great.

What about being a vegan or vegetarian? Be sure to read Mary Louise's last chapter to hear her exciting news about trying to clean up her diet and become a vegan. Although I may not have the level of self-control of Mary Louise, I do eat what I would consider a very clean diet. I eat very little sugar and meat, and not usually beef. Being a vegan might work for some people but not for me personally. I eat meat quite often, but not every day. I eat chicken; vegans don't. They don't drink milk or use other dairy products either. I drink milk once in a while and I also drink soy milk. I actually eat eggs once in a while and also cheese. Periodically I'll eat dairy products just to view life from a balanced point of view rather than from a fanatical point of view. Vegetarians are a lot healthier than the guys who are obese and thirty pounds or more overweight. You may just want to go to extremes and risk the danger of being called a health nut for a while until you get down to your ideal body weight. There are many other things to add to the list. You can go to the website and study detailed information about all these things. What I have to share with you is my own experience with many different types of nutritional plans and supplements that come from nature and modern science.

Juicing

Next, let's take a quick look at juicing. Juicing is simple and an excellent source of condensed nutrition. Jay Cordage, better known as "the juiceman," has gone around the country for years talking about juicing and how all the best stuff is found in the juice of any vegetable or fruit. He's a great advocate of carrot/apple juice and many, many other vegetable and fruit combination drinks. My own experience with juicing has been very positive. I juiced every day for seven years and I felt incredible every day and was able to function in life with full energy. I rarely ever needed a nap. The only thing I found negative is that I gained some weight during this time. It appears to be because I

What I've just described is "C.O.D," which means clear out, clean up, organize, and design. Your life is worth it. Do one space at a time. A briefcase or a purse can be one of those spaces. I guarantee if you have a briefcase so loaded with papers it throws your back out to carry it, you are overwhelmed with your work. If your briefcase dates back to World War II, I guarantee it doesn't represent the professional you are. Get a new briefcase and do a C.O.D. project on it and your whole life will open up. Try it and see for yourself. Here's a tip. Take home only what you are going to work on and complete that night. Here's another one. Go through every paper in your files, determine if it is still relevant in your life and work, determine under what category it would make sense to file it, and label the folder with a printed label. Do different colored file folders for different domains of your life, like "family," "personal," "clients," and so forth. It will be tedious, yet when it's complete just open your file drawers and feel how much more in control of your life you feel. Order brings quietude and ease. It also opens up space. When you can manage well the life you have, you can expand. It's worth the time. Go for it.

One of the physical spaces in which life occurs is the human body. If you are not responsible for the proper care and maintenance of the human body your opportunity for human existence will be cut short. It's kind of like gravity. It's really a bad joke too, that we came into this human existence without an operator's manual. You get a manual with a vacuum cleaner, but not for a human body? After trial-and-error experimentation for centuries now, we really have come up with a manual, so to speak, for the human body. Even though our knowledge about the functioning and care of the human body is greater than at any other time in the history of mankind, it is still probably minuscule mostly juiced carrots and apples, which according to this famous doctor's glycemic indexing, is high glycemic index. If you don't want that negative side effect, don't use high-sugar fruits or vegetables. I would go out and run up to 80, 90, 100 mile weeks, and I wasn't able to lose weight. Perhaps one of the reasons was that my metabolism changed when I was 23 and it will be more difficult with every decade. Bad news, but it's the truth. Anything that is high glycemic index creates fat. That's what you need to know. What I eventually did was cut down on the amount of carrot/apple juice I drank every day. I could have switched to celery/apple juice and reduced the sugar, but I like carrot juice so I decreased the amount. I used to drink 12 to 16 ounces of carrot/apple juice every day. After I found out how fattening it was, I cut that back to only four or five ounces a day! I didn't gain weight with that amount. I still got all the health benefits from juicing. I healed myself of many minor ailments through juicing. Juicing is an excellent tool for creating extraordinary health. On our website there are several juicing books from which to choose. We have also done the research for you on about eight different juicers to find the best juicer on the market. We like the juicers that don't produce pulp and the ones that clean most easily. After juicing every day you get tired of cleaning the machine and we found the very best machine. You can also find that information on our website.

There are some remarkable health claims about juicing. Recorded case histories show some people have overcome some forms of cancer and many other chronic debilitating diseases with juicing. Juicing has a great following and it is something we highly recommend on your journey to youthing. Juicing supplies your body with the enzymes and the nutrition you need to be healthy and to be on the road to a younger, stronger, and healthier you.

Many products are now being developed and made available to people that are nutritional breakthroughs. We want to share with you just some that have made the biggest impact for us. We searched for products in nature that were subjected to rigorous scientific research to prove their efficacy and safety. We discovered products out there that change lives, reverse the effects of normal aging, and produce extraordinary health. We will mention five of them here and briefly share our experience with each of them.

compared to what we don't know about this miraculous space suit that houses the human spirit's experience here on earth. Even though it is quite common knowledge, I'll go over the basics of what has kept me able to compete in a full-contact sport that requires balance, speed, endurance, and strength. What I probably value more than performance is looking 20 years younger than I am. It's a woman thing. You do know, don't you, that it's not common to speak of balance, speed, endurance, and strength, in the same breath with a 63-year-old. That I am the only one to have done what I've done in my sport may be proof enough of the validity of what has worked for me in maintaining my body strong and vitally alive. You will notice that I say "vitally alive" instead of healthy. Most people think of healthy as the absence of disease. That is too close to disease for me. I strive to be so far at the leading edge of healthy it needs another word to describe it — thus vitally alive. Then if I indulge in some of the goodies that are not good for us, I won't fall over the line into disease; I'll just have to recover some of my vitality. Please also notice the word maintenance here. It's a key word. Inherent to ownership is responsibility for maintenance. Responsibility is a word that has often gotten a bad reputation. To be responsible often feels "heavy," like we are in for some hard work. Worse, it sounds like we are to blame for something, as in taking responsibility. That is really an odd way to speak about responsibility. You can't "take responsibility" — we are always responsible. To acknowledge our responsibility seems a little like acknowledging that we have two hands or two feet. When we accept the truth of being responsible for the quality of our own lives, we regain our power to fully create anything of which we can dream...if we are willing to be responsible for doing the work to bring it into reality, no matter how long it takes. Darn, and I wanted it

Noni Juice

In Tahiti for many thousands of years, the natives have utilized noni juice. This is a very well known fruit that grows on a bush, not quite like a pineapple plant, but the noni fruit looks something like a pineapple. The juice of the noni fruit contains restorative phytonutrients. The key element that works in noni Juice is called Xeronine. Noni juice has been used by people all over the world for centuries and has worked miracles of health. One of the most incredible cases I remember hearing about was a little boy who had bone calcification, a disease in which your bones fuse together and normal mobility and function is lost. He drank noni juice for a few months and regained full movement of his hands and limbs. There are many other anecdotes of the health-promoting benefits of noni juice. In the Tahitian people who drank noni juice cancer was almost nonexistent. Pregnant women who drank noni juice had no morning sickness and enjoyed vibrant health during pregnancy. People experienced normal sexual function all their lives. noni juice is definitely worth trying. Drinking just one ounce of noni juice every day can change your life. Go to the website and click on the link for noni juice for more information.

Mangosteen Juice

Another juice has been popularized in the United States recently. It is the juice of the mangosteen fruit that is found in Southeast Asia. It is similar to noni juice in that its phytonutrients are restorative and healing. It has a far better taste though. They've been using the fruit of the mangosteen tree in Southeast Asia as a tonic and healing elixir for years. It is famous for restoring health. This product is being produced and shipped around the world right now. The sale and distribution of mangosteen juice is exploding. People are loving the effects of this elixir. Once again go to the website for more detailed information.

Mango™ juice is a nutritious drink whose main ingredient is the mangosteen fruit. Xango™ provides many benefits that will help meet your body's needs. It is the very first product on the market that uses the mangosteen fruit with all its *nutritional benefits*. Xango™ is taking the health food market by storm and is becoming the "it" product in nutritional supplements. Below are listed some of the published benefits.

all right now. Sorry, but bringing our dreams into reality with all the accompanying experience is how we grow and are refined in life.

If we have the blessing of owning something, we automatically have the privilege of the responsibility to maintain that thing. If we buy the car of our dreams and don't put oil in it at the prescribed intervals, even getting away with skipping a few of those intervals, we all know the outcome of that scenario. Of course, worse yet, what if the predictable outcome for that set of wheels was from putting sugar in the gas tank instead of gas? We also know the predictable outcome of the scenario of overeating, not exercising, taking harmful drugs, or smoking. So why do we think we can ignore the known outcome of damaging our human machine, much less ignoring commonly known maintenance regimens? Maybe it's really difficult to be responsible for abundance. We overindulge in what tastes good and feels good and then when our bodies are overweight, out-of-shape, and full of disease, we lose hope of being able to turn it around. We abdicate responsibility and say we are just getting old. If we give up responsibility, we give up the power to create a great body and life. I know you think you have no power over something like cancer, but I witnessed my husband create responsibility for having cancer as a self-determined experience in which he had lessons to learn. When he embraced this interpretation rather than being the "victim" of cancer, he had the power to discover what his body needed to heal itself, and he did. Thirteen years later he has no sign of the illness when he was supposed to be dead in a year. It may be too much for you to accept that we are responsible at that level and it may not be true; I just experience more miracles and power when I operate as if it is. I think it's always smart to use what works.

- Provides much needed and powerful antioxidants and health-promoting phytonutrients.
- Improves blood circulation.
- Strengthens the body's ability to fight disease and infection.

With *mangosteen as the main ingredient,* Xango™ contains absolutely no artificial flavoring, coloring or sweeteners. Xango™ juice is 100 percent natural and contains phytonutrients called *xanthenes* from the mangosteen fruit. xanthenes are some of nature's most powerful antioxidants. In Southeast Asia, mangosteen, because of its high xanthonecontent is considered to be the queen of fruits.

Xango™ juice is the only product in the world that contains this high amount of Xanthones and uses the mangosteen fruit in its most natural state. Xango™ is highly nutritious, it also tempts your taste buds, with a mix of strawberry, apple, grape, blueberry and various other mouth-watering fruits. All it takes is just one or two ounces of Xango™ juice in the morning.

Goji Juice

Goji juice comes from the fruit juice of the Lyceum plant. A neighbor introduced this product to me. There are people in the Himalayas that live to be more than 100 years old and still enjoy mental clarity, vibrant physical health, and vitality all their lives. Anthropologists have studied these people to figure out why they live so long and stay so well and vital. There is one claim that a man in the Himalayas lived to be two hundred and thirty-four years old. You can believe what you want, but a research team, including a famous American doctor and nutritionist traveled high into the Himalayan Mountains to a village known as Aureville, to study this phenomenon and to find out why these people lived so long. At first they thought it was the glacier water; then they thought it was this, or they thought it was that. What they finally uncovered was a famous little berry called a goji berry, rich in incredibly powerful antioxidants and phytonutrients. Many of the "phyto" or plant nutrients have never been isolated and identified, but their effect has been measured. The research surrounding goji juice has identified phenomenal human benefit. The research confirmed that the goji berry contains phytonutrients that restore health and vitality. goji juice is a powerful antioxidant. If you want more information on goji juice please visit our website.

Here is a body maintenance system that has worked for me. These are the basics. You need pure water. The optimum amount of pure water is about one gallon of pure water daily. That is much easier said than done. That is 16 eight-ounce glasses of pure water daily. Pure water doesn't come out of our taps, except if you live on the side of the mountain that I live on in Sundance, Utah. We have deep-well spring water coming out of our tap. Can you imagine such a luxury? Until we moved to Sundance I had a reverse osmosis water purifier. Reverse osmosis gets the job of purification done. There are other methods, but in considering them all, I found this gets the job done the best. You can study them for yourself if you like. Just drink pure water. What works for me is to carry a water bottle around wherever I go. I also recommend going to a health food store and buying the kind of plastic bottle that doesn't leach chemicals from the plastic into the water. In other words, don't use the water bottles you can commonly buy at the grocery store. The damage from the chemicals from that plastic negates the benefit of drinking pure water. Along the same lines of fuel for the body and what has kept me from the horror of middle-age spread, here are my dietary guidelines. For my body, proper food combining has been very important. When I hit 60, I gained 20 pounds, even exercising intensely for two to four hours a day and eating a better diet than probably 90 percent of all Americans. I'll admit that when I went to the movies, which was every weekend of my life, I ate a large bag of popcorn-by myself. I also only occasionally indulged in chocolate and rich desserts, but it was more then than now. I ate like this for the last 30 years of my life. Why would I start gaining weight now? In reality, every decade has been harder, but I was always able to cut out simple carbohydrates, cut down on amounts, and cut out snacking later in the evening and

Ascai Juice

Perhaps the most exciting of all the juice extracts is of the ascai berry. It has only recently made an impact in the world of natural health. This exotic berry was discovered in the Amazon jungle. The South American Indians have been drinking the ascai berry juice for centuries. It has been scientifically analyzed and is now considered by many to be a super food. It is a super-antioxidant. Amazon warriors drank the ascai berry juice before combat and it increased their strength and stamina. It is now used before sporting events for the same desired effect. In addition to improving strength and stamina, perhaps an even more desirable effect for both men and women is the enhancement of sexual desire and performance. Be sure to get further details on this amazing new discovery at www.secretsofthefountainofyouth.com.

Mistica Juice

A new product just released from Synergy Worldwide is a combination of all the most powerful juices in the world. This new super formula includes them all. The indigenous peoples of the Amazon call açai the "Tree of Life" and for centuries have used the tall, slender plant as a food and comforting balm for many health conditions.

Amazonian warriors reputedly drank the juice before battle to increase their stamina and strength, and local legends claim the juice has magical potency to enhance sexual desire. No other fruit can claim its extraordinary combination of antioxidants, amino acids, anthocyanins, and essential fatty acids. Synergy's Mistica captures this mystical energy in a high-powered fruit blend that fortifies your body with essential life-giving nutrients. Savory and delicious, Mistica gives you what you need to accomplish more each day.

We have all heard of free radicals and the negative effects they can have on our health. Free radicals occur everywhere in our environment, including in our food and in the air we breathe. The açai berry and the other fruits found in Mistica are natural sources of anthocyanins. An extremely potent type of antioxidant, anthocyanins are the pigments found in berries, red grapes, and other healthful foods. These compounds offer numerous health benefits, including

I would drop weight. At 60, nothing worked. I gained no matter what I did. I also know that cutting out too many calories slows metabolism further. So, I ate smaller amounts and ate more frequently to keep my metabolism higher. Nothing worked. Hopelessness was creeping in. It wasn't fair. With the amount of intense exercise I did every day, I should have been able to eat anything I wanted and still be skinny as a rail. Wah! Life is not fair. Call the "wahmbulance." Thank goodness that where there is a will, there is a way. Two things miraculously showed up in my life at that point. One was the product offered by the company Synergy Worldwide, called ProArgi-9. If you are interested in the full education on that product check our website for the appropriate link. You can also just take my word for it. Taken very specifically two hours after you eat on an empty stomach and thirty minutes before sleep, it stimulates your own body to produce the human growth hormone levels of a 16-year-old. They have hundreds of thousands of case studies over 20 years to substantiate this claim. It is safe, with no known side effects reported over a 20-year period. It works to reverse the commonly accepted effect of aging, called middle-age spread. Now...you also have to have a diet that supports maintaining an acceptable size and shape. You can't have a steady diet of Burger Biggy and have the body you want. It just handles this thing called a steadily declining level of human growth hormone with every decade of life. Without sufficient levels of human growth hormone, your body naturally puts on fat to handle your survival. Your body just wants to survive; it doesn't care that you look like 20 pounds of potatoes in a 10-pound sack. Many of the effects associated with aging are related to a lowered production of human growth hormone. ProArgi-9 has been a miracle in my life, emotionally as well as physically.

the ability to support the circulatory and cardiovascular systems. In addition to powerful antioxidants, açai contains essential fatty acids, amino acids, and plant sterols, which are compounds that resemble steroids in their anabolic effect with none of the negative side effects. Sterols help maintain healthy cholesterol levels that are already in the normal range. They support the immune system and are also useful in maintaining a healthy prostate gland.

Packed with tropical goodness, Mistica tastes great and provides your body systems with a dazzling array of health-enhancing properties. Including a daily serving of Mistica in your nutritional regime may help to enhance vitality and energy, while guarding against the ravages of free radical damage. Mistica is truly the nectar of the gods.

ProArgi-9 — The New Modern Miracle

I would say ProArgi-9 is the most important supplement of the 21st century.

We want to tell you about a recent discovery that is going to change the world. It is so powerful and such a well-kept secret hardly anybody has found out about it yet. Since the title of this book is *Secrets of the Fountain of Youth*, we are going to reveal to you perhaps the most important "secret" in its impact on anti aging and youthing. In other words, your quest to become younger and youthful will begin to be fulfilled from the minute you begin incorporating some of the products and methods we have suggested throughout *Secrets of the Fountain of Youth*. This tried and true protocol will positively and profoundly impact your health — mentally, physically, and emotionally. This ongoing, powerful, leading-edge research has been performed for over 20 years. The products are proven safe. These products are proven to be importantly effective to human health and performance. If it's in this book, we know its safe and we know it works. If you found something that was so powerful that it changed your life, wouldn't you want to share it with everybody? The secrets to youthing will now be in your hands and a part of your every day life. We hope you will share this book with everyone in your life. We bring that up in the book because we feel the product we are going to present next is the most scientifically sound, well researched, safe, and powerful product of

Basically eat no simple carbohydrates. Distinct from the Atkins diet plan, you do eat whole-grain complex starchy carbohydrates, just not with protein and fat. You can have veggies, which are an example of a complex fibrous carbohydrate, with brown rice or any other whole grain. I also bring my knowledge of glycemic index to the scenario and add barley to the brown rice, thus lowering the glycemic index of the rice. Glycemic index refers to the level of insulin produced in response to eating a particular food. Again, check our website for more information on glycemic index. Both white rice and brown rice are classified as medium glycemic index foods. Brown rice just has more nutrients than white rice, but the glycemic index is only one point different. By adding a low glycemic index grain, such as barley, you lower the overall glycemic index of the blend. Adding peanuts to whole grains also lowers the glycemic index, thus avoiding the insulin rush of higher glycemic foods. All you need to know about that is that a continual insulin rush will put fat on you, make your cells insulin resistant, and eventually wear out your ability to produce enough insulin, resulting in type 2 diabetes. That is not good. The problem with peanuts, which sadly enough used to be my favorite snack food, is that they are an improper food combination of protein, fat, and carbohydrate according to this theory. Peanuts are out. I eat fruit, but only alone. That's right, not even with yogurt, not even the nonfat, low-sugar variety. That is the end of strawberries and cream. There are some sacrifices, but you can still have rich, satisfying desserts, by following the simple food-combining guidelines. You can have cheesecake as long as you make it with a sugar substitute. I use Agave, a low glycemic index natural sweetener from the cactus plant. Do without the graham cracker crust. Some foods like peanuts should be avoided if you want to lose weight our time in the domain of health and anti aging. It has the possibility of making a bigger difference in your youthing than anything else we have discovered. It must be used correctly and specifically to get the effects you want. Pay close attention and believe what you are about to because it is supported by over 20 years of scientific research by some of the brightest minds in the United States.

I first came across this secret during my favorite hobby of running. I have been running marathons for over 20 years and do one or two marathons a year. In the fall of 2004, I contacted Mary Louise Zeller, who is the coauthor of this book, and told her I would be running in the St. George marathon in southern Utah. She got very excited and said, "You gotta try this new product; I don't have time to explain everything about it, but trust me it works!" She said she had been using the product for several months and it had truly altered her life. I told her I didn't usually try anything new that close to a race, but I had a great respect for Mary Louise. She is a world-class athlete and one of the most outgoing, trustworthy, and reliable persons I know. I knew she wouldn't give me anything that was illegal in sports competitions or dangerous to my health. Mary Louise told me the product was totally legal for athletes and was being used by many Olympic athletes. The product was a single amino acid that had been used for over 20 years with no known side effects by literally millions of people and was leading-edge technology for competitive athletes in training or before and during competition. Mary Louise assured me the product was 100 percent safe and there were zero side effects. She told me to just try the product.

I altered my course and went by her house and got some of the product to try at the marathon. I took the product just before I started my race and some at the 18 mile mark. I had amazing results. I actually ran the marathon faster than I had trained and discovered ProArgi-9 really worked. As I mentioned, I have been running marathons for over 20 years and I know my running ability and conditioning extremely well. I had trained moderately for about six months and felt if I had a perfect day I would run about 8 minute 30 second miles the whole way. That would have put my marathon time at about 3 hours 45 minutes. That's three hours and forty-five minutes. I actually ran 3:31 which is just barely over 8:00 minute miles. It may not seem like much to you, but to me it and can be added back in now and then on a maintenance plan. Some of the other foods to be avoided if you are trying to lose weight, are beets, carrots, corn, avocados, and olives. They are higher glycemic index vegetables or have the improper combination of carbohydrate, fat, and protein. I don't feel deprived or starving on this plan. I lost the 20 pounds. I turned around middle-age spread! I feel like I am in control. Without sugar in my diet, my cravings for sugar have disappeared. I feel self-disciplined, not deprived. I'm proud of my accomplishment. One of my own additions to this plan is to eat something raw at every meal. As we age, our bodies produce fewer digestive enzymes, thus being the source of much of the gastric distress commonly associated with aging. Live plants have live enzymes and will aid us in the digestion of food, eliminating that gastric distress. My body is thriving on this eating protocol and I have the energy level of the Energizer Bunny.

Now you are eating for vitality and performance and hydrating your body to a healthy level. You also get enough sleep for your body to regenerate itself and deal with all the toxins in today's toxic environment. That would be at least eight hours a night and women actually need more. Sleep is important to health and performance. Just do it!

Now that brings us to exercise. Oh no! That sounds like more hard work! Get over it and call it play. We used to, you know. We didn't dread roller skating, swimming, cowboys and Indians, kick-the-can or red rover. Don't call it exercise. Even if you lift weights, learn everything you can about it and enter competitions. Make it your game. Take ballroom dancing lessons and enter a competition. You'll see why you never see a fat ballroom dancer. You will want to do weight-bearing exercise two to three times per week. You need to do a game that includes aerobic (with oxygen), anaerobic (without oxygen), and isometric (muscle was nothing short of a miracle. I knew it was physically impossible for me to run that fast. Not only that, I usually die the last 6 to 8 miles and end up jogging in under very painful circumstances, wondering why I keep running marathons. I had a water bottle with the product in it at the 18 mile mark and couldn't believe that at the 20 mile mark I started feeling great. As a matter of fact nobody passed me after that point. I ended up passing hundreds of people over the last six miles. I ran the last mile in about six minutes. It was euphoric. I couldn't believe how great I felt. I finished as strong as I can ever remember. Needless to say, I really felt after the marathon that there was something to this product, so when I got done I asked Mary Louise about it. I got this explanation about the product that I must share with you and you must share it with the rest of the world. It is truly remarkable.

I was pretty skeptical in the beginning about what I was hearing because it sounded so unbelievable and it will to you too, until you actually go online and read the research and start using the product. Then you will believe! I must admit that at first, I just thought it was some sort of new energy drink and treated it that way for several months. What I discovered over time was it wasn't just a miraculous energy drink. There was something in this product that really changed the way my body was functioning. What I'm going to tell you in the following pages is so miraculous that we issue you a challenge. Try ProArgi-9! We know every single person who has put this product to the test has had a powerful response.

Before I tell you about the actual product, I want to express very briefly what the product has done for me. I have now been on this product for several months and some amazing things have occurred. I feel younger than I ever have before. First it has increased my energy level. It has increased everything you would want from a youthing point of view. You know all the things you think of when you think of youth. My libido has returned. Remember I am 57. I'm not afraid to say this, but my sex drive and everything else in my body has taken on a more youthful appearance and expression. I am able to do all the things I did when I was young. The results of taking this product are so phenomenal and are virtually undeniable. These are the things I want to say up front about the product. These are just a few of the things. Read on! Can you tell I'm excited?

toning through tension) movements. The martial arts fulfill all the above requirements, besides strengthening the spirit, building courage through competition, and providing self-defense skill, which in today's world is not a bad idea. The most important message here is to get your exercise by playing and mastering a game. You can have endurance, strength, speed, and balance all your life if you continue to exercise regularly and in an endeavor that requires endurance, strength, speed, and balance. Your body will give you what you ask of it. If you don't use it, you lose it. I'm 63 and I beat women half my age regularly. Ask any of my students what it's like to train with me. I can out-sit up, out-push-up, out-kick, and outrun 95 percent of my students who are in their teens and twenties. I'll let you know if I hit the wall, but so far the body is not the limit. I have a theory that people stop playing in life and then their bodies get old, not the other way around.

In addition to a diet that supports vitality and high performance, I use supplements. I know, I've heard the arguments that it's a waste of money, but I don't think so. I demand a lot from my body and I'm sensitive to what works and what doesn't. I've put together a supplement program from which I clearly experience positive results. I've experimented. I measured my performance with and without supplements. I perform dramatically better taking supplements. I take a high-quality, natural-source vitamin/mineral supplement, immune system support with antioxidants and herbs, circulatory system support with ProArgi-9 and Co-enzyme Q10, other cardiac formulations, mental-functioning-support supplements, and metabolism-stimulating supplements. Go to our website if you want to see an exact list of the supplement products I use. I am never sick and can physically perform like a 25-year-old. That is not normal and customary for a 63-year-old. Something is

I'm going to go through a description of why the product works and how it works and everything I currently know about the product so that you have an understanding of why this product and the research accompanying it, is one of the greatest breakthroughs of our time and the *master key* of the youthing process. What we want to deliver to you is the documented information that allows you to first believe in the product enough to try it. Then you will experience for yourself the miracles that come from becoming truly younger.

Let's just start off by naming the product and telling you that it was given to a company to market worldwide. The company is called Synergy Worldwide. They began marketing this product in the United States in February 2004 and you probably haven't heard about it yet or you would be taking it.

The name of the product is ProArgi-9. Let me tell you what it is. ProArgi-9 is a simple, single amino acid that was discovered many, many years ago but nobody knew how to deliver it in such a fashion that it would really make a difference in people's lives. We can acknowledge and thank the Japanese for inventing the technology of isolating and manufacturing individual amino acids. This famous research doctor, started researching this product as a result of her own health problems. You can go to our website and read about her whole life story; it's extraordinary. Let me just abbreviate it for the purpose of this writing. She was diagnosed with a very aggressive type of esophageal cancer that claimed 100 percent of its victims' lives. A famous doctor, who is most known for his vitamin C research, was then doing extensive research with L-arginine. At the request of her father, he agreed to try to save her life. He put her on large doses of L-arginine and a few other nutritional supplements. She was expected to die in a matter of weeks and instead she improved daily. Within a matter of months she was on her way to recovery. It is now over 20 years later and she has made a full recovery. She is a normal, healthy, functioning human being as a result of this phenomenal product and her own research. Because of her own dramatic recovery, she promised herself and God she would use the rest of her life doing research in this dramatic new technology. The biggest problem with L-arginine was it also produced some nasty side effects if taken continuously on a long term basis. It was her research that working. One of my goals is to participate in a scientific study in exactly measuring the improved performance from taking certain supplements.

Let's talk about another space that makes a difference. That is the space of integrity. Develop impeccable integrity, if you want to create space for miracles and extraordinary possibilities in your life. Most people first think of the most basic sense of integrity, which is honesty. It's true-you must first have honesty if you want your life to work. It's that simple. If you are not honest, that is, you lie, cheat, and steal, your life won't work. You may enjoy some success but you will not be truly happy or satisfied. One of the problems about being in that state of personal disloyalty is that you often lie to yourself about being dishonest. Things will get worse until you can honestly confront your own dishonesty. It may have to get very bad before that happens. Honesty is a basic of integrity. There can be no mastery, joy, making a difference, and so on, until you handle that one.

There is more to integrity than just being honest, though. The more you do what you say you will do, the more power you develop to bring your goals, dreams, and visions into reality. You literally strengthen the power of your word when you push yourself to fulfill what you said you would do. Sometimes you won't. You then acknowledge that you didn't do what you said you would do and make a new promise. It's a double-edged sword. If you are fulfilling everything you said you would do, you aren't making big enough promises. You just live from doing what you say you will do and watch your life take off in a direction I promise you will love. Another higher level of integrity is to be true to yourself and your ideals and values. First of all you must focus intensely to uncover the truth about the true self to which you are being truthful to and exactly what you do produced the product ProArgi-9, which after 20 years has no known side effects. She made the product safe.

This famous doctor is chief of biomedical research at the Glycemic Research Institute in Washington D.C. We talked a little bit about weight loss and glycemic indexing earlier in the book. She is one of the initial authors of this research and the chief technical researcher in the science of glycemic index. She is credible, she is amazing, and she has a list of references that will astound you. In her research of this amino acid, L-arginine, she discovered why it worked so well on her. She aspired to make the product into a safe and simple supplement. She had to make it bioavailable to the body so it could work its wonders. Twenty years of impeccable research has gone into producing this product. It is an exceptional nutriceutical now called ProArgi-9. It is very different from ordinary L-arginine isolate. ProArgi-9 was just granted a patent in the last few years because it is literally a proprietary formulation. There is no other safe L-arginine formulation that can cross the blood-brain barrier to have the same impact as ProArgi-9.

First of all, ProArgi-9 has been developed using impeccable scientific method. It is absolutely safe, it has been consumed by millions of people. About five or six million people have taken it without one single side effect! This is a safe product. When I first used ProArgi-9, I just trusted Mary Louise because I know her so well. But when I read the research it amazed me. It is 100 percent safe. As we mentioned ProArgi-9 has been consumed by five to six million people with no known negative side effects. That is outstanding. Put your mind at ease. ProArgi-9 is a safe product.

Secondly, is it proven effective? There are again millions of people who could attest to its benefits. It has been used by many competitive athletes and it is 100 percent legal as well as effective. Many Olympic medalists and professional athletes have been on this product with amazing results. A former, Mr. Universe, was on ProArgi-9 for over 14 years. A former, Miss Galaxy, was on the product for over 12 years. To our knowledge both of these professionals are still taking the product. These people are people who are in peak condition and they claim ProArgi-9 is the product that has enabled them to maintain that peak condition and level of performance even though they are aging and should be declining. They are recognized throughout the world for hold as true ideals and values. We know these things for ourselves if we take the time to inquire into it. Choose to live a life consistent with your ideals and values and watch your life give you more of what you really want. Integrity is the way of the warrior in whatever century you live.

I have proposed ways to create space in your life that will deliver to you more passion and aliveness. I've found it so and I share with you what has worked for me. I always say, "If it works, use it." Perhaps you'll try some of these ways to increase your experience of being fully alive.

their youthful appearance and extraordinary shape. Their success was not the result of illegal steroids; both of these athletes are committed to accomplishing extraordinary results only through natural means. ProArgi-9 is a single amino acid and is not harmful to the body in any way, shape, or form. It is simply a product that allows you to grow younger every day. It is considered to be a "magic bullet" for the cardiovascular system by the traditional medical society.

Here are just a few clinical things you need to be aware of about ProArgi-9 and about your body. When you reach the age of 23, you literally begin to die. Your body makes a shift from anabolic to catabolic metabolism. In other words, we are actually programmed for decline and extinction. The extinction doesn't bother me as much as the decline part. Wouldn't it be great to have a body that functions optimally and then one day you just die? It's the decline in function and breakdown that doesn't sound like fun to me. Does it sound like a grand reward for a life well-lived to get bald, fat, wrinkled, and not be able to do the things in life you have always loved? At the ripe old age of 23, the body slows its production of youthful hormones. It is the absence of these hormones that cause the aging process to accelerate. One of the hormones perhaps everybody knows about and was mentioned earlier in this book is called human growth hormone or HGH. Growth hormone is the hormone the body produces to maintain the function and appearance of youth. When you turn 23 the body responds to inherent programming to start cutting back the production of growth HGH. Every decade the production of HGH is diminished, causing the body to age and function to diminish more and more. Maybe God wanted to challenge us to take responsibility for the quality of our own lives by giving us this problem? The decline generally accepted as normal associated with aging can either be accelerated by poor nutrition and lack of exercise or it can be slowed down, significantly stalled, and even reversed by proper diet, rigorous daily exercise, and breakthrough supplementation.

As a perfect example of this, I remember at age 30 that something drastic happened to my wife. I'm sure the decline of the production of HGH had much to do with those changes, but I don't know scientifically exactly what happened. But something happened because around the age of thirty she got debilitating hay fever. She had never

NOTES

had it in her life. She'd never had any allergies and then all of a sudden she got hay fever. It was really strange that it happened right around that time of her life. You may have noticed the same thing. Around the time you hit your mid-twenties you had the same lifestyle, you were eating the same things, doing the same amount of exercise, and keeping your life absolutely the same and an interesting thing started happening. You started putting on inches and weight. This pattern is pervasive and everybody probably already knows this. It is no wonder this country is a third overweight and a third obese. It's commonly known and accepted when your body shifts into this catabolic phase from a reduced production of HGH you start aging and putting on weight. As part of that process you begin storing body fat and losing lean muscle mass. This is a scientific fact of life, a sad fact of life.

What if I told you there was a way your body could literally reverse that process, not just stop it, but reverse it. What if you could literally tell your body, "I am getting younger every day"? Sound familiar? That is, in reality, the message ProArgi-9 delivers to your body's aging mechanism. The body is always able to maintain the function and appearance of youth. It was just told to turn it off. ProArgi-9 can tell the body to turn it back on. You begin youthing rather than aging. As a matter of fact, if you stay on ProArgi-9, your body will continue youthing and return to functioning, almost as you did when you were sixteen years old. This is a scientific claim supported by research. This is worth further investigation. Can it be true? The only way to find out is to try it yourself, as we did. This is one of the biggest secrets of the fountain of youth that we have to share with you. Go to our website www.secretsofthefountainofyouth.com, click the link to ProArgi-9, and read the research for yourself.

On a rather comical note, I had a friend I shared this with and he said, "Well, I hope it doesn't take my brain back there because I was pretty stupid when I was sixteen years old." No, it doesn't take your brain back, you don't revert to your sixteen-year-old mentality or experience. Thank goodness, right? Okay, so that is the clinical quote of the century. Can you imagine? It is phenomenal! What we are talking about here is when you take ProArgi-9 at night, two hours on an empty stomach, and you take the amount that is recommended, ProArgi-9 actually crosses what is known as the "blood-brain barrier" and it stimulates your pituitary gland

NOTES

to start naturally producing HGH again. Isn't that a miracle of modern science? ProArgi-9 is a miracle. It's been tested by millions of people and has been proven to actually produce this effect.

This famous doctor has spent over 20 years in her scientific research developing a safe and effective product. Here is a list of the benefits and functions that are attributed to the extended use of ProArgi-9:

1. Enhances lean muscle mass.
2. Supports muscle growth.
3. Simulates muscle development.
4. Supports hypothalamic response.
5. Aids in decreasing body fat.
6. Is an antioxidant.
7. Helps maintain healthy blood sugar levels.
8. Is more well tolerated than L-arginine alone.
9. Boosts energy levels.
10. Is rejuvenative.
11. Helps promote healthy sex performance.
12. Is an adaptogen.
13. Helps provide the simulation and building blocks necessary for the body to maintain its own healthy growth hormone levels.
14. Helps produce nitric oxide (NO).
15. Stimulates production of human anti aging (youthing) mechanisms in persons over 23 years of age.

ProArgi-9 is truly the most potent nutriceutical ever discovered and has more benefits than any other nutriceutical product in the world. The government has actually said that ProArgi-9 is safer than any other L-arginine product available. It has been tested for over 20 years in millions of people.

What you have in essence is a product that will help you in every way in your youthing process. Truly it is the most powerful product on the planet. As a matter of fact, one of the greatest validations of this research is that the Nobel Prize was awarded for research on the nitric oxide production stimulated by L-arginine, and ProArgi-9 is the only safe and patented L-arginine product available. Nitric oxide is the biochemical that governs sexuality in men and women. ProArgi-9 increases blood flow by reducing and eventually eliminating plaque in your arteries. ProArgi-9 is the product of choice to increase sexual

CHAPTER 4

WINNING THE SECOND HALF

The Rest of Your Life Can Be off the Charts

Where does winning the second half happen? Isn't it at halftime in the locker room with a great coach? If you have a really good coach and you do what the coach says you'll have a better shot at winning the second half. Let's apply that to life. If the first half is the first forty years, how have you done? Mostly it's been trial and error, hopefully learning from our mistakes. Very few of us had the advantage of a life coach. Going back to our analogy of competitive athletes, training is fundamental to success. Training can also be fundamental to our success in life. Sometimes we function in both males and females. The Nobel Prize awarded in 1998 was awarded for the research defining the effects of L-arginine on nitric oxide production. It's a great validation of what we have been saying.

You must take two scoops of ProArgi-9 at night, thirty minutes before you go to sleep. It must be taken on an empty stomach in order to cross the blood-brain barrier. You can also enhance lean muscle mass development, physical stamina, and performance if you take one scoop of ProArgi-9 about thirty minutes before you exercise.

We have just shared with you perhaps the most powerful nutriceutical that will literally launch you into orbit in your youthing journey. What more can we say? Your own body turns off the aging process and turns on the mechanisms of youth. It is a scientific breakthrough in the technology of youthing. There isn't anything more we can say about this phenomenal product. This is the single most powerful anti aging, and the most powerful youthing tool on the planet today. Everybody over the age of 23 should be taking ProArgi-9. Even if you are young, it will keep you young. If you are 88, ProArgi-9 has the same effect on you as on someone who is 28. It will stimulate your own body to naturally produce everything it needs to function as if you were sixteen years old again! Isn't that spectacular? Everything you need to know about ProArgi-9 is on our website. There you'll find many testimonials about its breakthrough benefits. ProArgi-9 is one of the most important keys to unlock the secrets of the fountain of youth.

The 15 benefits associated with the use of ProArgi-9 are legal health claims! That's what took 20 years. The Food and Drug Administration has approved those claims as legal health claims! Isn't it amazing almost every supplement you pick up in a health food store carries a statement that reads something like this, "This statement has not been evaluated by the Federal Food and Drug Administration"? Well, the label on ProArgi-9 reads just the opposite. These are fifteen legal claims related to the patent granted to this research doctor. The actual patent # is 6001809, if you want to do further "due diligence" or research on your own.

ProArgi-9, unlike any other L-arginine formulation, makes those claims and backs them up with twenty years of documented and scrutinized scientific research. Substantiating her research, when she filed for the patent, she submitted over 750,000 scientific papers to are well trained and well prepared for the game of life and it follows that we would have success. Maybe you are not well trained, yet you played well and had unpredictable success. You also could have had minimal training and preparation for life and your results are commensurate.

As we enter our senior years, even if we experienced more than average success, I dare say only a few of us are working and performing at our best, inventing new paradigms of reality for ourselves, and leaving the legacy we always dreamed. All too often our lives have been used up paying the mortgage and supporting our families. We often are too busy and too spent at the end of the day to dream any more. We sometimes lose heart and still go through the motions of living our lives, even if we are successful and producing a great income. Probably still fewer of us are doing work about which we are truly passionate and in which we are whole-heartedly engaged.

By the time we are entering the second half of life most of us are pretty comfortable by the world's standards. We go out to eat two or three times a week, go to the movies, have engaging athletic endeavors, enjoy our friends and family, go on some great vacations, and so forth. Especially if we are comfortable, we seldom stop and think, "Am I doing my dreams in life? Am I contributing to the quality of life for all people? Am I inventing new possibilities for all human beings?" Did we just stumble into our jobs or did we diligently plan and work toward the life we created? Sometimes our work is fulfilling and we are contributing to mankind, yet we still are unfulfilled in our closest relationships. Can you imagine creating and planning a life that is a masterpiece? If you don't plan for a masterpiece you will have a future that is predictable and consistent with some version of your past. In our country that is usually more than OK–and yet many of us feel our the Food and Drug Administration.

ProArgi-9 is used by thousands of world-class athletes, including the NFL, PGA, Mr. Universe, Ms Galaxy, Triathletes, U.S. Olympic Teams, and IBFF. Over a twenty-year period, Synergy's ProArgi-9 is the only arginine formula that has not had one negative side effect, not in one person, after use in over 6 million people.

Including ProArgi-9 in your youthing supplementation program, will give you cardiovascular benefits, anti aging benefits, normalization of blood pressure and cholesterol benefits, normalization of blood sugar benefits, an increase in lean muscle mass, and a decrease in body fat and all of this from one innocent little amino acid.

There is no apparent end to the realm of benefits of ProArgi-9!!! You should not wait another second. Go to the website and find out how to order this unbelievable product.

This is youthing in a can.

At the printing of this book Synergy had just released a new product as a spin-off to ProArgi-9. It is called Encode and has some major benefits to particularly, the minority population. Go to our website for the latest information on this exciting new release!

lives are somehow still unfulfilled and incomplete. We often have many accomplishments with which we are truly satisfied and yet we know that we have greatness within us, real contributions we could make, dreams, and goals that somehow still remain unexpressed. What's the block that stands in the way of full self-expression? We often attribute our tiredness and apathy to getting older. Where did our passion, wonder, curiosity, excitement, and energy go? Passion, excitement, and energy are by-products of creation, full engagement, and full self-expression. We know our health and energy are not expanding and we have often gotten so out of control with our bodies that we have given up and succumbed to what I call the all-pervasive myth of the decline of aging.

What is this myth of aging? Don't we expect to get weak, have aches and pains, be inflexible, bald, fat, wrinkled, unable to perform, have decreased mental function, and no energy to do the things we have always done and enjoyed? When we began spending all our time working, we often just stopped playing. We then told ourselves, "We are just getting older and can't do what we once could." The truth is that when you stop playing hard you get weaker. When you again play and exercise, your body will strengthen. The body is not the limit! Don't we live in fear, hoping one of life's major illnesses won't suddenly strike us? We think it's just the big lottery in the sky that determines if it strikes us. We think we are helpless to determine our own health. As we enter our 60s, 70s, and 80s aren't we discounted and often considered to be a liability, even a burden? We literally become castaways. This is a grand reward for a lifetime of experience and contribution. I don't know about you, but I've decided I am not going to participate in the commonly accepted myth of the decline of aging. I will say, though, that unless you are engaged in strengthening and generating

CHAPTER 8

THE SKIN

The Magic Key To Youthful Appearance

What does the skin, especially the facial skin, have to do with aging? Why is wrinkled and sagging skin associated with the aging human body? Let's review a little basic biology about the skin. The skin is the largest organ of the body. One of the main functions of the skin is protection. The skin protects our inner tissues and actually holds our body together. The skin helps the body maintain the right temperature and it is the organ of touch. The skin is one of the most vital parts of the body and we need to take care it.

There are essentially three layers of skin. There is the outer layer called the epidermis, the layer just underneath the epidermis called the dermis, and the third layer located below the dermis called the subcutaneous layer or fat layer. I will very simply and briefly explain how the skin functions. If you just look at your skin you really can't tell what's going on, but actually there are some interesting things happening that are not visible to the human eye. New skin cells are being produced every second of every day. Cells move through a process of being created, aging, and dying. New epithelial cells are formed at the level of the enduring health and well-being, you will get the myth by default. In this commonly accepted culture of aging you have a future that is "on rails." You will have a future based on what you know about aging and how others see you as you age within this culture.

The nasty truth about this scenario is that in many ways we see getting old as the easy way out. We are literally co-conspirators in creating this culture. "I've worked hard all my life; let them take care of me now. It's my turn to take it easy." We sometimes see aging as an effective tool to control others. What are some of the other benefits of going along with the current culture of aging? How about being comfortable, having no responsibility, and having life continue to be safe and predictable? In this paradigm we are resigned to our own interpretations of who we are that are lies about ourselves we learned from the past-from our parents, from our teachers, our friends, our bosses, and even from groups who hardly know us. We learned to believe what we told ourselves about ourselves or believed what others told us. We often lived our entire lives with no idea of who we really are and what we are capable of. The only time we are really living is right now. The only problem is that our past not only colors our experience of right now, it is the only future we can imagine for ourselves. Real power is created in actual moments of now.

The access to power is to have moments of now untouched by the past. We can generate a whole new possibility in the future, if we are free from past-based illusions and reactions. In order to complete the past you must first understand that it does not exist…really, other than in language. Even subconscious thoughts are held only in language. The access to completion is in language, the basic element of communication. Then instead of a future of more, better, different versions of how you have been "basement membrane" which separates the epidermis from the dermis. As they age, these cells move toward the surface layer of the epidermis, finally dying and sloughing off. The process, from when your skin cells are formed until they die and are shed, takes anywhere from 14 to 30 days. Cell formation, maturation, and death are repeated over and over until we die. Every day you are actually creating, shedding, and recreating over 40 million skin cells.

Your epidermis, or outer layer, constitutes only 5 percent of the total volume of skin. The amount of pigment, or melanin, is what determines the color of your skin. Beneath the epidermis is the dermis layer, which is obviously unseen because it's covered with your epidermis. The dermis is basically a manufacturing plant. The blood vessels are located in the dermis layer of the skin. Sweat glands and nerve endings are also found in the dermis. The oil glands that keep your skin pliable and moist are located in the dermis. The dermis produces several elements that keep your skin looking young, firm, and elastic. The two most important elements essential to this function are collagen and elastin and we are going to explain how they function.

The blood vessels found in the dermis are the tiny capillaries that carry all the life-sustaining nutrients to your skin. These capillaries carry oxygen and nutrients to the skin cells and also carry away waste material. The oil glands keep your skin lubricated and the oil protects your skin. These oil glands are also called sebaceous glands and at any given time they are producing what is called sebum or natural skin oil. These oils keep your skin lubricated and flexible, preventing the absorption of too much water through the skin.

You have sweat glands in the dermis as well. You are sweating all the time, 24/7. Sweat is excreted through tiny openings called pores. When the sweat mixes with the sebum, the protective coating that preserves the skin is formed.

You have nerve endings throughout the dermis. The nerve endings give you your sense of touch or feel. It is really important that the skin allows you to know whether things are hot or cold. The nerve endings send impulses to the brain every split second and the brain distinguishes whether something is hot or cold. Those nerve endings not only sense temperature, but if you touch something, the brain interprets feeling being up until now, you can generate a future in which you are free to be who you create yourself to be and are willing to take the action to align your dreams with reality. A freely created future is a powerful future that reaches back and gives us who we are in the present moment. Think about that for a moment. Imagine someone called and told you that you won the lottery. Your windfall, as delineated on your written contract, would be 12 million dollars. You would receive your first check on the first of the month and 1 million every month for twelve months. You wouldn't have the money in your pocket now, but can you see it would definitely impact who you are being now? Just knowing of the incredible future awaiting you at the first of the month, you would be an entirely new person.

Can you see that this is how human beings actually are? Knowing this gives us a chance to profoundly alter the possibility of our lives. The question is then how to create a future of greatness that reaches back and gives you a new way of being now. If you want to break through to greatness, it's smart to get a great coach. My husband, Ron Zeller, is just such a coach. This is the work he has done for thirty years with people all over the world, from many different social and political cultures. My husband and I developed a two-and-a-half-day life-altering course called "The Fenix Event ...a course for winning the second half ." Please go to our website www.secretsofthefountainofyouth.com and look for "The Fenix Event" and see if this is the arena in which you will create a whole new relationship to age and aging and a new possibility for your life. This book talks about it. The course is the doing of it. There may be other paths to the same goal. This is just one that I know works. Our lives are clear examples of this technology realized. Ron and I continue to expand the possibility of our lives when most people our age are slowing down and taking it easy.

and distinguishes whether something is sharp, whether or not it hurts, or whether it's dull or pressure. All of those senses happen through nerve endings. Nerve endings distinguish feeling if you touch something. Nerve impulses go to the brain, which responds by interpreting the sensation.

The third layer, the subcutaneous layer, is beneath the dermis and is made up mostly of fat. The epidermis and dermis are both relatively thin layers of skin. The fat layer is thicker to keep your body warm and to absorb shock. About the only thing you'll find in the subcutaneous layer other than fat, is tiny little tubes called hair follicles. These tubes line the hair and take it up through the dermis to the epidermis. You have these little follicles all over your body except on the palms of your hands, soles of your feet, or on your lips. These follicles are also the vent for the subcutaneous glands. The subcutaneous glands secrete oil to maintain the elasticity of the skin. Working from this basic knowledge of the skin, we can now have a great conversation about youthing as it relates to your skin.

The skin, particularly the facial area, is one of the key visible areas that reveal how the body is aging or youthing. Aging of the body takes place through two major influences. The first influencing factor is the level of human growth hormone production. Diminishing of the levels of HGH affect how you feel, your musculature, your ratio of fat to lean muscle mass, all functions of the body, and the quality and integrity of the skin. All these effects are governed by your brain and the production of HGH. The biggest barrier to shifting your body into youthing mode is altering the body's natural programming to diminish HGH production with every year of age. The lack of HGH is mostly responsible for all the effects of aging. As production of HGH diminishes, the skin also reflects significant signs of aging. Another major cause of aging effects of the skin is exposure to interior and exterior toxins. The skin, being the body's external protection system, is particularly injured by environmental toxins and prolonged exposure to ultraviolet rays of the sun. An example of interior toxins are those from cigarette smoking or chemicals ingested with processed food, producing millions of free radicals and other pollutants. Both the inward and outward aging of the body are accelerated by the shift to catabolic metabolism and by the accompanying slowing of HGH production.

If you have been slowing down and taking it easy, have you noticed it's not very satisfying? I think it seems like you are just marking time, even if you are enjoying yourself, until you die. I invite you into a new world of possibility. Please join us in Sundance, Utah for "Winning the Second Half Summit." I promise you it will be a pivotal point in your life no matter what your age.

Our lives are often used up to just get by, even if we get by really well. When we get old and cannot produce what we once did, we are of no use and discounted. We even discount ourselves. It is a mass conspiracy! If you could raise your own head up out of the culture long enough to see it and if you were interested in generating a different possibility in aging, how would you create it?

You would first examine your life and the culture in which you live and discover where you are and what is really so. If you look closely enough and long enough you will see a clear distinction between a reactive life and a created one. You have just taken the first step in generating a new reality. No reality can be imposed on us without our agreement. Sometimes we give our agreement by offering no disagreement and by creating nothing. We can either create something or create nothing, but we are always creating. To make a difference in the quality of life for others and ourselves, we must first own our power to create. Notice I didn't say we must gain the power to create. We are always creating, even if we are creating agreement with something that has already been created. We are creative beings; it is elementally who we are. We sometimes create consciously, sometimes unconsciously, or create by our agreement, but we are creators. The quality of life for all human beings has always depended on the human beings who accept the challenge and responsibility to consciously and committedly create that which has never been, that

As the skin ages, the common visible signs are the appearance of lines, wrinkles, and sagging of the skin. Another phenomenon that shows up in your face with aging is puffiness and dark circles under your eyes. Age spots, liver spots, and other discolorations of the skin are all signs the skin is aging. The skin also begins to thin and lose its elasticity. We start looking old. We all recognize those signs and can guess a person's relative age by the appearance of their skin. The question is, what can be done to alter the effects of aging commonly accepted as normal — the lines, wrinkles, sagging, age spots and discoloration of the skin. Can youthing of the skin actually take place? For example, can you cause your skin to look younger just by taking ProArgi-9? ProArgi-9 can cause the rest of your body to feel and perform like that of a much younger person. What about the skin?

Taking ProArgi-9 for at least three months stimulates your body to produce the hormone levels of a young person, along with the resultant return of youthful function and performance. That is a clinically proven statement and is an absolute breakthrough in the science and technology of human aging. Reversing the effects of the aging of the skin and repairing damage to the skin is very difficult. Until now the modern methods that have been developed to deal with the aging of the skin are wounding the skin with dermal abrasions, surgical face lifts, chemical peels, and chemical injections. Ouch! ProArgi-9 can produce dramatic changes in your bodily functions, your sex drive, your musculature, your weight, your body fat, your energy level, the healthiness of your organs, and it can even improve the way you feel. It is one of the great secrets of the fountain of youth but perhaps the most difficult problem is to reverse the damage and aging of the skin. Even though ProArgi-9 aids your skin along with the rest of your body in looking and feeling younger, Synergy Worlwide has also pioneered specific dermatologic products to bring about youthing of the skin.

Synergy Solutions is a line of skin care products that are a scientific breakthrough in what is known as signaling technology. A bipyramidal liposome delivers an enzyme to the dermis layer of the skin. This enzyme signals to the fibrocytes to again start producing the kind and amount of collagen it produced when you were young. Collagen is the scaffolding structure that supports the skin. When collagen is diminished, the which is now not commonly accepted, or that which is not now known. All breakthroughs have depended on daring to think the unthinkable!

Isn't it unthinkable to desire to compete in the Olympics in a sport that requires strength, speed, power, and endurance when you are in your 60s? It is even more unthinkable to train like that was really possible. Yet it has been engaging in that impossible dream that has strengthened me beyond what people can believe for a person my age. I have become quite a novelty. People ooh and ah when I lift my leg so high I bounce my thigh off my chest and my flexed foot actually goes behind my head. I can wow them by dropping into a full split without even stretching out first. I can blow their minds by popping a target so hard and fast it sounds like a rifle shot. A 63-year-old can't have that kind of speed, power, and flexibility; everyone knows that! That is why I am invited to do keynote talks. I begin with a demonstration of the martial arts. Then they listen to everything I have to say about reversing the effects of aging, about nutrition and wellness, about doing the impossible, or about anything! Anyone can do it, anyone who will create it and do the actions to produce it. I don't have any special gene for rare strength and health. It's not an accident of genes. I just didn't believe the myth; it didn't look like much fun to me. It's that way with everything in life. At any time, at any age, and under any circumstance, you can transform the quality of your life. Since we can say anything and cause a reality, why not say something that empowers us to greatness? You can do anything you say you can do if you are willing to do the work to bring it into reality, no matter how long it takes. What else are you doing for the rest of your life? Have some real fun. Do what hasn't been done.

skin sags and wrinkles. These fibrocytes always could have continued producing this collagen, but they are literally programmed to stop producing collagen and start the aging process earlier discussed. When signaled to begin producing collagen again, they do. These products were developed by the laboratory that discovered an enzyme found in certain sea plankton. This sea plankton enzyme very closely resembles a human enzyme that is fundamental in protecting the skin from ultraviolet damage from the sun. This enzyme prevents damage to the DNA strands at the cellular level. This technology was fundamental in saving the lives of the "bubble children" and was incorporated into a skin care line of products that does indeed reverse the effects of aging and damage to the skin. Mary Louise says this skin care line has made her skin look and feel 10 to 15 years younger. She says that after using the products for a year the lines around her mouth, eyes, and on her forehead have regressed 200 percent. Noticeable improvement showed up after only one month of use. Science is advancing, and reversing the effects of aging and youthing of the skin is now a real possibility without expensive and painful surgical or chemical procedures! There are so many products with so many claims, that one of the biggest problems is trying to identify products that truly can alter your appearance and reverse the aging process. We have used hundreds of products and are presenting those that have made a difference in our lives and truly can do the job for you. Each person is different, but we believe that our years of research and effort will pay off for you. That is why we wrote this book.

Obviously, ProArgi-9, good nutrition, and exercise all help your body to function better and generate the younger you. You could actually start turning around the effects of aging by just taking ProArgi-9, exercising regularly, and altering your diet. If you included more enzymes in your diet, either through juicing or eating fresh fruits and vegetables, you could begin to reverse the effects of aging of your body and your skin immediately. "Start youthing today" is our slogan. Do what works. As a matter of fact, you could be younger tomorrow by just eating fresh fruits and vegetables for the whole day, today. If you started exercising today, you would start youthing as well — getting younger is a daily choice. If you can run 5 miles in less than an hour at age 45, and a young man at the age 18 can't even run 20 feet without

Now here's one of those secrets of the fountain of youth. There is a door to a quantum shift in expanding who you are, your experience of life, and what you can bring forth powerfully in life. I'll label the door "Mastery of Space." We talked about the power of being responsible for the physical space of your environment. There are other spaces in which to engage and master that will produce a quantum shift in the quality of your life. One is the space of integrity, another is your spiritual space, your financial space, your space suit (your body), emotional space, the space of relationship, the space of communication, the space of responsibility, and the space of action. There are others, but when you take on mastery in these spaces, your life will blast off into the miraculous zone. You can't even dream or imagine what is possible in this domain. Herein lies satisfaction and joy in life. Engaging in mastery in these domains will spawn the fulfillment of your legacy in life. On this path you won't discover who you are, you will generate a new possibility of you.

We have talked about integrity as having the power of your word. In the basic sense it's being honest. That is your ticket into a real life. Another part of building strong integrity is keeping your word or communicating beforehand that you won't. Just like daily practice of kicks makes for very strong kicks, the practice of telling the truth and doing what you say you will do strengthens the power of your word. One of the things I had to deal with in that domain was recognizing that I was a "pleaser." If my husband asked me if I would do something I always said yes. I often did not keep my word and I had lots of reasons why I didn't. We humans often don't know that reasons are not the same as results. We think if we give a good enough reason why we didn't do what we said we would, it's as acceptable as the result. Consider that, in stopping and huffing and puffing with exhaustion — who is younger? If you can perform any event at a higher level than a younger person — ask yourself who is the younger person? Aging is relative to how you look and how you feel and how you perform, not the number of years you have spent on the planet. You are actually youthing; you are not aging, if you are taking ProArgi-9, other nutritionals, juicing, and getting exercise every day. These are all essential elements of youthing. But, can you speed up the process?

Common sense tells us healthy skin will have a better appearance. This is so much common sense it might not seem like a secret of the fountain of youth, but each key is important in unlocking the door to youthful, smooth, beautiful skin. We recommend an exact program to follow based on our own experience and on the new science and technology of skin care. We will give you access to all the products and information that will change what you think is possible about the youthing process of your skin.

When someone looks at you to guess your age, they don't look at your arm or any other part of your body, they look at your face to make that determination. Therefore, we will focus on what most people don't know about their face and how to restore youth to your face and the rest of your skin. There are two ways to help the skin start youthing. There are internal treatments and external treatments for the skin. We will start with the inside.

Internal Skin Treatment

In evaluating the impact on healthy skin from the inside out, consider, "You are what you eat and drink." We need to eliminate anything we ingest that is damaging to our skin. We need to ingest anything that is good for your skin. First, consider what you drink every day. As we mentioned previously, acidity is not only bad for your whole body, it is devastating to your skin. You want to eliminate alcohol, coffee, any caffeinated beverages, soft drinks, orange juice and anything other acidic drink from your diet. Sugar and chemical sugar substitutes are also enemies to your skin. This includes soda pop, diet sodas, or any kind of fruit juices that aren't 100 percent fresh extracted fruit juices. Most of the time you are going to be drinking, you guessed it, good ole life, we either have results or reasons why not. Of the two, results make a difference in the quality of our lives and reasons do not. Results are satisfying and reasons are not. It's more honest to say we didn't keep our word and forget the reasons why not. Then we can either recommit ourselves or withdraw our commitment to do it. I was downright passive-aggressive about saying yes to everything, wanting people to think I was some kind of superwoman and a really good and capable person, and then resenting that I had to do it and just not doing it. That personal pattern is sure to drive the important people in your life crazy. Your family loves you and will probably keep tolerating it, but bosses will eventually fire you and your friends will gossip about you. Life is so much easier when we say what is really true for ourselves, rather than saying what we think others want to hear. Now sometimes we have to do things we don't want to do. Mothers do the responsible thing in taking care of their families, not just what they want to do. I'm not saying just do what you want to do. With some requests we really do have a choice. If we give in against what is really true for ourselves, we will get them for it!

We often do not stop to consider what our true self-expression is. We often react to what presents itself in our lives or do what others think would be good for us. I remember that I went to nursing school because my mother and my grandmother always wanted me to be a nurse. I made straight A's and I hated nursing. I was miserable in nursing school. It didn't light me up; it was someone's good idea for me. My husband has been an extraordinary influence in my life. He coached me to look into myself at great length and depth until I pulled out of myself what I loved, what truly lit me up. It seemed as though I had lost myself somewhere along the line. There wasn't much natural self-expression left by the time I met him when water. You must drink adequate amounts of water. How much you need to drink depends on the amount of exercise you do and where you live. If you live in a dry climate or a wet climate, and whether the climate in your area is defined as hot or cool determines how much water you need to drink. The seasons of the year are not as important. It doesn't matter whether it is winter, spring, summer or fall — year-round you want to be drinking at least two to three quarts of water every day. You can drink other liquids as well. Teas are good, especially herbal teas. Despite all the hoopla about the antioxidant qualities of green tea, it's not the best thing in the world for your skin. Freshly prepared fruit juices are great just don't over do it with the concentrated natural sugar in fruit juice. Again, to get all the nutrients and fiber fruit has to offer, it is always better to eat the whole fruit. One of the drinks Mary Louise enjoys when she goes out to restaurants is lemonade, but not the lemonade they prepare. She asks for water and a few extra lemon slices, which the server is always happy to provide. She squeezes the lemons, whips out her Stevia Extract (a natural low glycemic index sweetener, found in a health food store) from her purse, and drops in a few drops and viola', — lemonade! You need more water if you do lots of physical activities or exercise. You obviously need more water if you live in a climate that is warm or dry. You do need more water in summer than in winter, though in winter you need adequate amounts of water to produce body heat. You need more water, as well as more natural fiber, if you have digestive problems such as constipation. As a side note, if you do have problems with excessive diarrhea or constipation, psyllium husks are the perfect addition to your supplement regimen.

You actually can get by with drinking less water if you have a diet that is high in fresh fruits and vegetables, because they contain their own water.

In order to have a healthy body and healthy skin, there are lots of other things you want to avoid, as you probably have guessed. Avoid over exposure to the sun. Even short exposures, without applying sunscreen to protect your skin from the damaging ultraviolet rays are very destructive to the skin. Although there are recent studies which claim that the literature about the sun being bad for you is false. Sunshine is a good source of Vitamin D that the body readily accepts.

I was 38 years old. I had a good job. It was interesting and I was proud of myself for doing it. I made very good money, but I haven't a clue how I ended up doing what I was doing. I think I spent more time picking out a new dress than choosing what work I would do or which man I would marry. Martial arts light me up. Competition lights me up. Teaching lights me up. Standing for the champion in everyone lights me up. I did nothing for three days but question into who I really was as a true self-expression in life and committed to generate myself to be that possibility. I have been living that possibility for the last 16 years. Very few people will have anything more than a reactive life without doing that work. It takes working with great coaches to have unpredictable, great results in life. So, in the ultimate sense, integrity is also being true to yourself.

Sometimes we have a long string of incompletions in life, either with people, events, issues, or getting things done. I suggest making a list of everything you can think of when you ask yourself the question, "What's incomplete?" Along with identifying your incompletions, write down possible actions for completion or resolution. Strike them off the list when they are complete. You will see that your list is finite. You can complete it. You will also recognize a new flow of power in your life as you complete each item. Powerful results are definitely tied to integrity. So are joy, peace, and satisfaction. If you are stuck in life, make an integrity list. Complete it. Then do another one. You'll discover a whole new level; you'll see things you couldn't see until you scraped off the first layer.

The space of spirituality is an integrity issue. Remember you don't have to believe anything I say. You can even argue against it. I'm just sharing with you what has worked in my life to have success, results, relationship, love, satisfaction, and exciting opportunities beyond my wildest dreams, and

Vitamin D has been shown to prevent many types of cancer. Research has revealed Sunshine Vitamin D helps reduce the risk for colon cancer, and there are also many other cancers that are inhibited by Vitamin D. That means you have to get adequate sunlight. What we don't want is to go out and do things that damage the skin. Don't go out in the sun for hours on end and get sunburned, for example. You know the ultraviolet rays do damage your skin. You be responsible for that too. When we are talking about getting enough sun we are talking about fifteen minutes at a time, maybe four or five times a week. Be very careful not to get overexposure from the ultraviolet rays of the sun.

The other thing that damages your skin is smoking. Obviously, if you smoke you are generating tons of free radicals and it does extreme damage to your skin. I know you have seen the skin of longtime smokers and have noticed the hardened, yellowish look of their skin. Think about what smoking does to internal tissues and organs. The outside effects are certainly a clue. Smoking even damages collagen production and thus a smoker's skin is even more wrinkled, especially around the mouth. We highly suggest if you do smoke, to quit immediately, if not sooner.

The same food you eat to produce a healthy body will help you produce healthy skin as well. The wisest principle to follow is to eat food as close to the natural state in which it grows as possible. Avoid processed foods. They are dead. How could they give life to your body? What you consume has a very powerful influence on your skin. There are foods that will internally help your skin be softer, firmer, and more youthful. It is not a great secret to know we should eat lots of fruits and vegetables. We highly recommend you eat up to five fresh fruits a day. Fruit is necessary for your body to produce soft, youthful skin. Lots of raw vegetables are also beneficial to your skin. Salads or lightly steamed vegetables are great. The darker and the greener the vegetable — the better they are for your body and your skin.

There is also a whole set of other nutrients that help the skin stay soft and pliable. These nutrients are essential for beautiful skin texture and elasticity. Skin vitality necessitates consuming of fatty acids. Essential fatty acids are found in vegetable oils, seeds, and nuts and you need an adequate amount of all these. Olive oil is excellent always being fully and vitally alive. You can discover for yourself that God exists. It's one of His promises to us. He said that if we want to know Him and the truth regarding faith in the unseen, all we had to do was inquire of Him with a true desire and faith that we would receive an answer. God is His word. I have discovered this truth for myself over and over. When I read scriptures, I pray and then I sit quietly with my eyes closed. I just sit and listen for a still small voice. Sometimes I even plead with God to talk louder because I'm such a dullard. He does too. Many people pray, but then don't hang out and listen. Communication is speaking and listening. Listening is a very important part of prayer. Some people call it meditation. Meditation is listening, but usually without prayer. With prayer it is communing with God. He will have something to say if you listen along with speaking to Him with faith.

This is a little story about faith. When we first bought our home in Sundance, Utah, we still hadn't sold our other home in the valley. Five years before, we bought a piece of property halfway between where we then lived in Orem, Utah and Salt Lake City. We had plans drawn for a beautiful home for this property, but we never started to build. The truth was that we were being practical when we chose to build there. Ron would be closer to the airport. We rationalized that it was up high and had a view of the whole valley in front and the mountains behind. The truth was that we loved Sundance. We hiked there several times a week. We loved the feeling of Sundance. We just loved being there. Sundance gets a lot of snow in the winter, though. It would be difficult to deal with the snow. We thought we should be practical. We finally told the truth. We really wanted to live in Sundance. We started looking at homes in Sundance and found the most beautiful, unique house I have ever seen. It sits on a sunny knoll with a full to use in cooking. Olive oil is impervious to heat so it can be used for any kind of cooking. It also has a very desirable taste and can be used on salads or for dipping bread. Olive oil is a great source of essential fatty acids and for the many nutrients that feed ageless looking skin. You should know that recently eggs have again been given a good bill of health. For many years we ate eggs every day. They were considered to be very healthy. Then researchers told us eggs had cholesterol in them and were bad for you. People took it to heart and quit eating eggs altogether. I didn't listen to the propaganda, because eating eggs helped my running. I've always eaten eggs. Regardless of the research, I knew eggs could not do too much damage if I ate them in moderation. Now the experts are saying the amount of good that is done by eggs, far out-weighs the bad. Eggs are good for you and you can start eating them again. We have permission from the scientific community to eat eggs again. Isn't that interesting?

Fresh fish is also a good source of essential fatty acids and many other nutrients. Fish should be included regularly in your diet. Remember, your digestive tract is the processing plant for all the raw elements your body needs for beauty as well as health and extraordinary performance. Give your body the raw material it needs to produce outstanding health and performance. The skin just reflects the job you are doing on inner health. Eventually what you eat shows up in the quality of your skin. If you are going to get rid of toxins, you need to flush out your body with lots of water. Adequate water is extremely important. Dehydration also shows up in your skin. There are other things that can help your body maintain the skin as well. You need a good source of protein. Protein is essential to the skin to maintain the constant growth of new skin cells.

Earlier we mentioned ProArgi-9 as an essential supplement for extraordinary health and youthing. ProArgi-9 also helps your skin to be healthy and strong. Of course, an increase in human growth hormone and nitric oxide will produce a more youthful skin. Supplements are critical in an aging body and multivitamins and minerals are excellent for your skin. On www.secretsofthefountainofyouth.com we list a whole group of vitamins, minerals, and other nutritionals that can help your skin look young again. The type of minerals and vitamins are important.

view of Mount Timpanogos, with waterfalls cascading down the front face of the mountain. It stands surrounded by wilderness only 15 minutes from town. It feels like I'm eternally on vacation and living in a postcard. Sure it's harder dealing with the snow, but even the winter has an incredible beauty and essence in Sundance. We have pure air and pure spring water flowing from our tap. What more could you ask for?

Do you remember that I told you that we still hadn't sold our other properties? That means we had three property payments all of a sudden. Ron worked hard to get the house in Sundance. It was a good deal, below market value. We got it. Then reality struck. We had the home of our dreams, but Ron was getting almost panicky realizing how much more money he would have to produce. We thought our house in the valley would sell quickly. It didn't. Our lot was the last view lot in the custom home section that hadn't been built upon. We thought it would sell quickly. It didn't. We thought we would move most of our furniture up to Sundance and rent it out for the Sundance Film Festival. Many of the homes like ours rent for $1,500 a night! Ours didn't rent and we spent additional thousands decorating. It was at this point Ron knew that with all his experience and power he was insufficient. This was the biggest, most stressful money situation he ever created and he thought he might have made a big mistake. He began praying. Now Ron does everything intensely. He prayed daily for two hours at a time. He prayed and prayed. I often walked into the bedroom and found him on his knees praying. He kept this up for weeks. He's got intention and stamina I have to say. It was during this time that our son Adam said, "Why don't we just go ahead and move everything up to Sundance and live there?" It was like a bolt of lightning for me. I just saw it was the thing to do. It felt like climbing

Both Mary Louise and I use the leading-edge formulation from Synergy Worldwide. This formulation is complete and encapsulated for bioavailability. You should include a full range of minerals and vitamins in your supplementation program and they are presented specifically in a chart and diagram found on our website.

Feeding our bodies the proper nutrients and drinking a lot of pure water are essential to having healthy, ageless skin and are truly secrets of the fountain of youth for the internal treatment of your skin. For the total picture; go to www.secretsofthefountainofyouth.com

External Skin Treatment

The other essential element of skin care is the external treatment. As we grow older without intervening with leading-edge technology, by default our skin will reflect the decline associated with age. Usually wrinkles, lines, liver or age spots, skin discolorations, and lesions begin to show up. These are associated with aging, but should we relegate ourselves to accepting these signs of aging or can we do something about it? Help is here! Some amazing research has been done over the last 10 years creating a breakthrough in skin care products and other things that will bring back your youthful appearance without surgery. We are going to mention a few of them. Collectively, we have tried them all and have once again discovered those that really work. Here are the latest and greatest secrets for youthful skin.

Sleep

Sleep is one of the other things we should mention before we present the products and how the skin responds to different external applications. Whether you call sleep an external or internal element is irrelevant. Adequate sleep is essential to the external appearance of your skin. Perhaps one of the most important and critical areas for rejuvenation and detoxification for the skin comes in the form of sleep. Most people have a tendency to not get enough sleep. It seems the older you get, the less sleep you get. Some people think that less sleep is good, when it actually is very harmful. Some people even brag that they don't require much sleep. Somehow they believe they have elevated themselves beyond human requirements. In order for you to out on skinny branches and holding on for dear life, but I knew it was the thing to do. The real estate agent said that houses don't sell as well if they are completely empty, but I still knew it was the thing to do.

Ron was working all the time and much of the time out of town. I began moving everything up to Sundance. Looking back on those two months, I don't know how I did it. Even with a lot of help from my friends and hiring some help, I still felt like a pack mule. I packed up the house, loaded the SUV and made two or three trips a day up the mountain, in February in the heaviest snow in 20 years. I unpacked and settled every load as I brought it. I contracted the small, unfinished cabin off the deck at the new house to be turned into a beautiful office for Ron. I spent more money on new office furniture. I shopped for new carpeting, lighting, and other decor. I had the kitchen floor refinished. By the way, don't ever refinish a wood floor after you have cleaned the whole house impeccably, because then you get to clean the whole house impeccably again. At the same time I told Ron that I knew he didn't want to hear this, but I needed another $5,000 to do some upgrades in the old house so it could compete with a new subdivision a few blocks away. I knew that was what it would take to sell it and get our price. Ron trusted my intuition, gave me the money and kept working and praying as hard as he could. I'm proud of myself that I did so much for so little in that house. The money seemed to stretch further and further like the seven loaves of bread and a few fishes. With that $5,000 I bought all new brushed stainless steel appliances. I even changed all the cabinet pulls to match. It took about 10 hours to change those cabinet pulls. It was hard work. The dad of one of my students does construction. I hired him to do work in both places. I don't know how he did it all. I shopped for materials and he did the work. He detoxify and rid the body of the toxins that affect your entire biological system and your skin, you need a minimum of eight hours of sleep each night. One of the essential elements for overall health and youthing, is getting adequate sleep. You should start making the changes in your life necessary for you to get at least eight hours of sleep every night. If your skin is going to look young again you must get adequate sleep and apply all the internal and external things we are delivering to you. We have talked about some natural protocols to follow — things the average person hasn't thought about and therefore doesn't do. The entire protocol is what makes the difference. Make sure you exercise, use the products, drink lots of water and get eight hours of sleep every night if you want your youthfulness to return

External Products

This is some of the most exciting and fun things we will address in this book. Imagine you can actually have your youthful appearance return. There are so many products and bold claims out there that it is difficult to sift through and decide what to even try. We have done that for you. We have found the actual topical formulations you can apply to your skin to help correct the years of damage that has accumulated, particularly on your face. There are literally hundreds of skin care products on the market and even more articles written about skin care. There are regimens such as exfoliation, or scrubbing old, dead cells off your face with a really good facial scrub. There are all sorts of oils and natural and chemical products that can be used to treat your skin. We are not going to spend time here talking about those things. If you want total information on any of these things the greatest website in the world is called www.secretsofthefountainofyouth.com. It has everything you need to know about skin care. On the website you will find all the latest research and many different products for skin care.

The purpose of this book is to reveal to you the secrets we personally have found that actually work. Many new products have come from cosmetic manufacturers and from powerful companies that market through network marketing. I'm sure you have heard of many of these products that have been designed to slow or reverse the effects of aging on the skin. The research over the last ten years has been a technological breakthrough put in new stone tile floors in the bathrooms and laundry room. He painted. He put up new light fixtures. I even had new hardware put on the doors and new carpeting in the family room. Even paying him as well as purchasing all the materials, I still spent only $5,000. That $5,000 kept going and going and going. The day the work was completed, it sold. We had that house for sale for a year. That day it sold. That was a big relief from the financial pressure on Ron.

I kept creating space in the new house. If you read the chapter on creating space you will have an idea of what I was doing. I was cleaning, organizing, and designing. It was a new level for me in designing and organizing physical space. I had miracles show up everywhere. Again, I did so much with so little. I don't know how I did it. This home is glorious. It's a whole new level of life for us and is literally changing us. All manner of new opportunities have shown up. We have now sold our other property. We made almost $100,000 on that property. Nothing usually sells in the dead of winter up there, but ours did. Given we live in a famous mountain ski resort that does nothing but go up in value, a friend offered to be a 10 percent partner in the Sundance house for another $120,000. It seemed like a miracle to us. It all started moving when Ron started praying. It had just been stuck. Ron is clear that he learned an important lesson from things occurring the way they did. I'm clear we acted on faith and that God helps those who help themselves. Over and over again God has answered my prayers and I testify to you that He does exist. We are not alone. If you bring yourself to Him, He will make Himself known to you too. I wish miracles for you too. Having God giving me guidance and direction has definitely been the source of extraordinary power in my life. Going to the next level in life and bringing your dreams into reality is in this domain. There are so many products out there it is confusing. We are going to simplify all that. We are recommending a line of products that will simply generate the youthing of your skin. All these products are thoroughly tested, thoroughly proven, and personally tested. We are going to hand you the information you need to begin using a few products that will allow your skin to fully recover and restore its youthful appearance.

There are many good topical treatments that simply hide or in someway temporarily improve the appearance of your skin. Some of these things are useful as cosmetics. In other words, they only cover up the evidences of aging. A couple of these products you can find on our website; we won't mention them here by name. These are products you can put on your face and cover up the wrinkles and skin discolorations. One of the products actually does away with the shadows on your face. When you use cosmetics you are not really doing away with your wrinkles, lines or age spots, all you are doing is masking them. They've actually developed a light reflection technology skin cream. When somebody looks at you, they can't see your wrinkles. It's amazing. If you put it on half of your face and go look in the mirror, the half of your face you put it on will look five, ten, fifteen, or 20 years younger. Just by putting this product on you will look younger and all it does is reflect light so people can't see your lines and wrinkles. Go to the website if you are interested in this product and look under facial products at www.secretsofthefountainofyouth.com.

The second product we have tried and recommend, if you want to actually look younger without having to go through months of traumatic and expensive processes, is also a topical treatment. It will make your skin look younger immediately. If you are impatient and want something to make you look better during the months that it will take to do it right, use this. This product tightens your skin and pulls your wrinkles tighter without surgery. It can make the bags under your eyes and the wrinkles literally disappear — temporarily. This cosmetic treatment takes an hour. If you want an instant look and want to do something right away and something that will work in the interim until the good stuff takes hold permanently, use this. Once again go to our website www.secretsofthefountainofyouth.com. You can use the quick index to find all these products on our website.

not an easy transition by its very nature. If you want more out of life, you have to risk it...and demand it with your intention, your trust, and your work.

What I say about the space of relationship is that there is work to do. Could you imagine buying a business and thinking it would run itself and everything would turn out great? We get married and think we will be eternally happy. We think if we married the right one we wouldn't have any real problems to work out. After all, we are in love. When you are in love, it just works out. Right! The only ones who still think that aren't married yet. It takes daily diligence and hard work for a body to develop skill and strength; that same daily diligence and hard work is needed for a relationship to develop. There is no staying the same. There is no real "hovering." Even hummingbirds are beating their wings hundreds of times a second in order to stay in the same place. If you are not growing, you are dying. There is nothing in this world that doesn't contract if it's not growing. Think about it. Does grass stay the same? If it's not growing, it's dying. If a relationship is not growing, it's dying. It takes daily commitment over time, a lot of time, in the project of relationship to have mastery in the area of relationship. It doesn't just happen! You have to do the work. There is no champion without the work. You say you don't know how to do the work of relationship. Get a great coach and practice. We get a manual with a vacuum cleaner, but we just get married and "have at it." There is "stuff" to learn about relationship. Get to work. If you want to learn a new language, or a new anything, you get a great teacher, find some good resources, study, and practice.

After a lot of the above, you discover you have some mastery and then you learn more by teaching others. There is no "It's done." This is "schoolhouse earth." It's always

Now we can get to the products for really youthing and transforming your skin. We want everyone who tries these products to take before and after pictures so that you don't ever forget the wonderful youthing transformation that is going to take place. We are going to introduce a whole line of products to you by name. Coincidentally, some of them come from the same company as ProArgi-9. The reason is simple. This company is really dedicated to youthing. However, all the products we present are dedicated to you and your youthing journey. This is not an advertisement for one company. In my opinion, Synergy Worldwide has many products that are leading-edge technology in the supplementation field, but these two sets of products, ProArgi-9 and their skin care line called, Synergy Solutions, are scientifically advanced products that are here ahead of their time. These products represent many years of scientific research. They're safe and they work. ProArgi-9 was researched for 20 years and clinically proven to cross the blood-brain barrier, when taken very specifically, stimulating your own pituitary gland to again produce the human growth hormone levels of when you were young. Their skin care line is also the result of years of clinical research.

This skin care line includes all cutting-edge products specifically designed to reclaim your youthful skin. They have two categories of products called Gen3 and Gen4. We have included only the products we have tried and believe in. You will see why we believe in the products when you begin your own skin care program. These products will literally transform your appearance – naturally.

Gen3 products represent one of the latest advancements in youthing technology. The skin care products that have been developed and the anti aging processes that have surfaced in the last few years are truly miraculous. The technology will *do* what cosmetics have only promised to do in the past. Even if a good cosmetic line promises to reduce the appearance of fine lines and wrinkles, it is only temporary. Transformation of your skin is when all the layers of your skin begin youthing. Gen3 products will dramatically and visibly reduce the appearance of fine lines and wrinkles — permanently. I imagine if you stopped using Gen3 or Gen4 products, over time your skin would begin to age again. The way most cosmetics work is if you don't use them that very day, your wrinkles are visible the next day. Most cosmetics just soften lines or cover them about learning. If you stop learning you're dead. You could say the same thing about the body and moving. In fact, I will. If you stop moving, you're dead! Here's another important "duh!" about marriage. If you want a marriage that lasts forever, don't get divorced. You'll have to start over again. Now, I have to add that there are some basics in behavior that are prerequisites to requirements to use that last statement. If you are in an abusive relationship, get out. If you are in a relationship with a person who continually betrays the trust of a marriage relationship, get out. You know when you are being damaged in a relationship. Get out. Especially get out if the children are being hurt. You don't necessarily have to get divorced. You could pray a lot for the other one. I know a man who married a very beautiful woman. You know that can sometimes be a curse as well as a blessing. She was constantly unfaithful and eventually asked for a divorce. She got the divorce even though he never agreed to it. She traveled all over Europe with rich men. Being very beautiful can get you some goodies in life.

Meanwhile back at the ranch, my friend still wore his wedding ring, he never went out with another woman, and he didn't "get on with his life." He lived his life; he was just still married. He truly held no malice in his mind or in his heart for her. He knew she was just where she was. He was committed to their marriage even though she wasn't. She divorced him, but he didn't divorce her. Now, I don't know if it would always work out this way if just one stays committed to the relationship, but after five years of this trial of commitment, he got a phone call in the middle of the night. It was his wife (for her, his ex-wife). She realized he was the love of her life, that everything she experienced over the last five years was a meaningless folly, and that she wanted to be married to him, if he would have her. He up. The Gen3 and Gen4 products stimulate the production of collagen again so the lines and wrinkles fill in. They're not there any more. It takes about one month of use to see a significant change. Mary Louise has been using them now for a year and a half and her skin has continued to improve. Someone guessed her to be 38 years old just recently.

I have actually been using all these products along with Mary Louise. Who says men aren't vain? For an actual regimen of how we use these products daily, go to the last chapter of my section and the last chapter of Mary Louise's section for an exact listing of the products we use and how we use them. You will get a play-by-play of every single thing we do in our lives each day to keep us young. Here I will simply list the products in the order I use them and you will get the company's description of the product and its benefits. Each of these products is cutting-edge technology for your youthing.

Synergy Solutions

Renewing Facial Scrub

An important step in optimal skin care is to periodically give your face a deep, thorough cleaning. Synergy's Renewing Facial Scrub is the perfect formulation for this delicate task. This advanced-cleansing product washes away makeup, oil, and pore-blocking impurities without harming sensitive skin. Pure jojoba micro-beads soothe and cleanse, releasing your skin's natural radiance and giving it a luminous glow.

Renewing Facial Scrub uses sulfate-free surfactants to thoroughly clean your skin and leave it refreshed. It won't dry out delicate skin or damage its ability to keep water in and allergens out. Jojoba micro-beads are naturally soft and cushioned. They gently exfoliate the skin, unlike other products with apricot kernels or walnut husks that cause tiny lacerations and irritate the skin's surface.

Renewing Facial Scrub contains generous amounts of sea-salt minerals, similar to the body's natural mineral content, to nourish your skin with essential trace elements. Natural botanical extracts such as grape seed, ginseng, and olive leaf keep your complexion moist and supple and provide antioxidant protection. With Renewing Facial Scrub, your skin gets the special treatment it deserves to stay soft, silky, and youthful.

caught the first plane out. They were remarried. They did some of the same Landmark Education courses I did. They did some counseling. They did more seminars to discover more about themselves, each other and how to structure a relationship to work. That was 20 years ago and they are still happily married. Like Ron and I, they say they are more in love with each passing year. There really is stuff to learn and stuff to practice to have a great relationship. Get to work.

There is also stuff to learn about the space of communication. There is the common model of communication that I will call reactive communication. All that is said is said in reaction to what they said. I'll call it "Newtonian communication." One of Newton's theories was that a body at rest would remain at rest until acted upon by another body in motion. It's kind of like billiard balls. One ball hits another ball and sends it flying. That's like communication in this system of communication. We throw words at each other, impacting each other, starting another round. Communication in this domain is an attempt to justify yourself, be right, protect yourself from a real or imagined verbal attack, control, and dominate. Did you ever notice it doesn't work? Look around. It'll be the majority of communication you'll hear. In this system of communication, one "thing" is trying to get another "thing" to see a certain point of view, do what this thing wants that thing to do, protect this thing from that thing, and so on. Have you ever noticed there's no affinity in this system of communication?

Another possible model of communication would be authentically creative and "re-creative" communication. I'll call this system of communication "quantum communication." As in quantum physics, it is the space between things that is of primary importance, not the

Delicate Facial Cleanser

This extra-mild, pH-balanced cleanser contains powerful antioxidants, nutritive vitamins, and botanicals to leave your skin feeling clean and refreshed. Unlike many skin products, it contains no harmful ingredients or harsh chemicals that can leave an unsightly residue on your skin. You'll love this product's silky texture and pleasing aroma the moment you apply it. Delicate Facial Cleanser is formulated with the latest liposome technology to deliver nature's best ingredients to your epidermis and leave it feeling radiant and renewed. Use it for just a few days, and your appearance will radiate youthfulness, vigor, and health. Follow with Revitalizing Toner.

Revitalizing Toner

Revitalizing Toner is the final step to cleansing. It gently sweeps away pore-blocking impurities, restores surface pH and prepares your skin for moisturization. Revitalizing Toner is a wonderful second step to your daily skin care regimen. This gentle pH balanced toner gently removes the last traces of makeup and cleanser and leaves the skin feeling cool and refreshed. It helps remove all residual dead skin cells and other impurities, while re-introducing unique moisturizing ingredients to your skin. Fortified with plant extracts and antioxidants, this Revitalizing Toner soothes and balances your skin, tightens pores and reduces the appearance of fine lines and wrinkles. Revitalizing Toner includes an exclusive blend of the latest free-radical fighting vitamin and botanical technology. The base contains Witch Hazel distillate for its astringent properties and the revolutionary long-lasting moisturizer Bio saccharide gum-1.eshener. Follow with Anti-Aging Serum.

DNA Repair Creme

DNA Repair Crème is part of the next generation (GEN4) of technology to use special enzymes and biotechnology to minimize the visible signs of aging, promote vibrant, youthful skin, and restore and revitalize your appearance.

Many factors, such as pollution, sun exposure, and poor diet, can affect skin condition. Your body naturally produces reparative enzymes, but as the body ages, it produces fewer of these protective agents. To things in the space. You can actually create space in communication. You can generate communication as a creative act, not as a reaction to what they did or said. You can truly listen to his communication and see it from his point of view and accept it as his point of view, without reacting and defending. You can honor his point of view without agreeing with it. This takes work. It takes education and practice. It does exist, though, as a real and viable domain of mastery in communication. In the common culture of communication the result is separateness. In the self-generative, causative, creative and re-creative domain of communication, the result is love, affinity, mutual respect, and unpredictable results. Get a great coach. Do the work.

In all of these spaces, engage in a course of mastery. It may be the road less traveled, but it is the road to being fully alive while we are alive. Ron Zeller has worked with people in the mastery of these spaces for the last 30 years in cultures all over the world. Again, I invite you to join us in this work of having the rest of your life be the best of your life. Having a great life is one possible way to live life. It may not be the "right" way to live life. It's just the best experience of being alive. This is our work in The Fenix Event.

support optimal epidermal health, DNA Repair Crème contains the T4 Endonuclease V enzyme. This enzyme is derived through a bio-fermentation process from a certain type of bacterium, and placed in a liposomal delivery system designed to reach areas of the skin where it will have the most beneficial impact. Once released, DNA Repair Crème's powerful ingredients help unlock the youthful appearance you have been missing.

DNA Repair Crème combines in one system the most effective ingredients and most up-to-date technologies. Used as directed, it will help bring back your skin's youthful tone, appearance, and texture. It is like no other product on the market.

TriAction Age Defense Complex

For ages, men and women have dreamed of discovering the fountain of youth. So great is the appeal of a youthful appearance that many people have resorted to drastic measures such as chemical peels, microdermabrasion, and collagen or Botox® injections. But these inconvenient, costly procedures can cause painful side effects to your face and wallet.

The discovery of advanced-action ingredients and the development of an exclusive skin-care formulation have led to an exciting new product called TriAction Age Defense Complex. A true wrinkle-reducing solution, TriAction uses a unique combination of powerful ingredients to achieve and maintain younger-looking skin.

The key to TriAction's revolutionary effects lies in its proprietary blend of peptides, plant extracts, and other natural ingredients. Peptides are potent biochemical agents that function as chemical messengers in the body. With age, collagen and elastin production decrease, causing wrinkles to appear as the skin loses moisture and firmness. A great deal of research has recently been done investigating the role of peptides in building the skin's collagen levels and relaxing muscle contractions, both of which help diminish the size of wrinkles.

TriAction combines three peptides (acetyl hexapeptide-3, palmitoyl oligopeptide, and palmitoyl tetrapeptide-3) to regenerate your skin and counter the visible effects of aging. Each peptide compound performs a different task in aiding you to look your best. Acetyl hexapeptide-3 helps

NOTES

to noticeably decrease the depth and size of wrinkles, while palmitoyl oligopeptide and palmitoyl tetrapeptide-3 promotes increased skin thickness and optimal epidermal balance and tone.

Along with the peptides, TriAction also includes Kampuchea, or Black Tea Ferment. Known as "long life fungus" in Russia, Kampuchea confers a number of positive benefits on the body, especially for the skin. It supports increased irrigation and circulation, leading to improved elasticity and texture. As volume is restored to targeted zones in the skin, the appearance of visible wrinkles is noticeably reduced, and your skin will look smooth and soft. Kampuchea also improves color and tone, while imparting a vibrant, rosy glow.

With TriAction Age Defense Complex, a fresh, youthful appearance can be yours, without the pain and expense of aggressive, professional treatments.

Advanced Face Cream

Advanced Face Cream incorporates sophisticated Gen3 technology to diminish the visible signs of aging and help give your skin its optimum moisture balance. Advanced Face Cream is quickly absorbed and leaves the skin soft, radiant, and resilient. Gen3 technology and ingredients represent the most recent advancements in skin care that work in harmony with the body's cellular processes. The Gen3 ingredient complex includes a combination of antioxidants, plant extracts, vitamins, and other pure ingredients that help your skin look younger and more vibrant.

Packed with lipids and essential fatty acids, Advanced Face Cream replenishes the skin and gives it the protection and moisture it needs. This light cream intensively moisturizes without clogging pores or leaving behind a heavy, greasy feeling. Your skin will feel clean, fresh, and new.

Advanced Face Cream is also fortified with a unique blend of the latest free-radical fighting vitamins and botanicals. Included in this blend are vitamins A, B, C, and E, aloe vera, Centella asiatica extract, Japanese green tea extract, Ginkgo biloba extract, and Echinacea purpurea extract. Biosaccharide gum-1, a skin-condition agent derived from sorbitol, soothes and moisturizes.

NOTES

Anti-Aging Serum

Environmental forces, stress, and the aging process can extract a grievous toll on the skin. As the years pass, the production of collagen and elastin decreases and the skin becomes thinner and less elastic. Sunlight can also break down collagen and elastin and cause damage of its own, including discoloration, blotching, and other irregularities. As a result, firm, taut skin begins to sag and wrinkle, and the youthful appearance is lost.

Synergy Worldwide has created a unique formula to support your skin's ability to resist these destructive forces. Anti-Aging Serum, with Gen3 technology, can visibly reduce the appearance of fine lines and wrinkles and revitalize how you look.

The power of Gen3 technology lies in its unique complex of ingredients and its revolutionary delivery system. The formulation includes dipalmitoyl hydroxyproline, a botanical amino acid that rejuvenates your appearance and fights against the cumulative effects of aging. It also assists in scavenging free radicals and helps moisturize, smooth, and firm the skin.

Gen3's ingredients are delivered by specialized liposomes, which are microscopic vesicles with a distinctive shape. This complex design is essential to properly release the ingredient complex to the necessary areas of the skin. In addition, the Gen3 liposome delivers a special "signaling molecule" that adds to its anti aging properties.

Anti-Aging Serum is also fortified with a unique blend of free-radical fighting vitamins and botanicals, including vitamins A, C, and E, Aloe barbadensis, Centella asiatica extract, Japanese green tea extract, Ginkgo biloba extract, grape seed extract, and Echinacea purpurea extract. Biosaccharide gum-1, a skin-conditioning agent derived from sorbitol, is added to provide long-lasting hydration.

There are many other products we have tried and are still using. Our personal research and ongoing product testing results will be published in a weekly, Secrets of the Fountain of Youth newsletter. Make sure you go to our website and sign up. It's free!

There you have it. The most powerful, technologically advanced line of facial products on the planet and they work. This line of skin care products uses the latest technology that actually has your own

CHAPTER 5

MY OPTIMUM HEALTH AND PERFORMANCE

Take a Sip from My Personal Fountain of Youth

If you want to insult me, tell me I am able to do what I do at my age because I have good genes. That I look like I am in my 40s instead of my 60s or that I have the flexibility, strength, speed, and stamina of a 30-year-old is also not an accident of good genetics. People actually say that to me when we meet, and I invariably answer, "I beg your pardon. What I can do at my age is the result of years of hard work, research, consistent training, and engaging in living correct principles."

fibrocytes producing collagen again. The Gen4 technology of DNA Repair Cream will help your skin repair itself and your youthful appearance will be restored. You will be amazed at what will happen to your face and the other parts of your body as you use these powerful, effective, leading-edge products for youthing. Please know that this is not about selling you products. We are giving you access to the newest technology and information — assembled all in one location. We have been your lab rats with these products for the last 10 years. We have saved you thousands of dollars. As you combine the regimens and the products of *Secrets of the Fountain of Youth*, you will reverse the aging process and look and feel younger than you ever thought possible. Remember each chapter contains secrets and what works best is to go back and master each area and make it your own. These things worked for us and I know they will work for you. Even though you think you know or understand a principle or product, until you gain mastery it will not show up powerfully in your life. Collect all the keys we have given you and unlock the magic door to true youthing.

Let's start with nutrition. You don't have to have a PhD in nutrition to figure this out. Keep it simple. Eat plant-based foods. Eat food as close to the state in which it grows as possible. I eat a lot of vegetables. I eat something raw at every meal. The older you get, the fewer enzymes your body produces. The enzymes in the raw food help your digestion. I also take an enzyme supplement. I eat fruit; I don't drink fruit juice. It has too much sugar and without the fiber, causes an immediate insulin rush. Treat yourself now and then to a sweet, but otherwise eat foods with a low glycemic index. In other words, regularly eat as little sugar as you can. You can get all this information on our website and educate yourself about which foods are low glycemic index. The glycemic index of a food relates to how much insulin is produced in response to eating that food. Foods that have a high glycemic index stimulate the body to produce more insulin. Over time, the constant presence of insulin in the bloodstream produces insulin-resistant cells, stimulating the production of more insulin. Too much insulin makes you fat. Sugar makes you fat. It can also over-stimulate the body's ability to produce insulin until it breaks the system down completely, which is called diabetes. That's all you need to know about that.

Eat whole grains. Americans are often overfed and undernourished. Whole grains are simply more nutritious. Organically grown whole grains are preferable. If our bodies get the nutrition they need, we tend not to crave simple carbohydrates or overeat. Brown rice is a whole grain and very nutritious, yet it is a medium glycemic index food. If you mix whole-grain barley, which is a low glycemic index grain, with brown rice, you lower the glycemic index of the brown rice. You could mix peanuts with the brown rice and accomplish the same result. If you read about glycemic index, you'll learn how to incorporate this very important

CHAPTER 9

THE PLAN

Putting It All Together

We need to put everything together into a holistic plan of action. You need to create a plan that will put you back in the driver's seat of your own life. This plan will be so powerful you cannot fail. Your job is to take all the secrets we have revealed to you and make them your own in such a way you are able to powerfully transform your life. Once you were aging and possibly aging in every single area of life — mentally, physically, emotionally and spiritually. You can turn off that old aging process and turn on the youthing. First of all, turn on your brain and get it to put all of this together in a powerful way that allows you to be fully alive. Begin the journey to youthing and stay on this journey for the rest of your life. You are now going to put together a plan to actually generate the new youthful you. The journey begins immediately, if you haven't already begun.

principle into your nutrition program. Try some of the more exotic high, protein grains, like quinoa. Beans, nuts, and seeds added to whole grains make complete proteins.

When I was younger I was an omnivore. Now I mostly eat vegan. I'm not "religious" about it. If I go out to eat I can usually find a good vegetarian selection, but I have been to a meal in someone's home and instead of disappointing them by not eating what they prepared, I will eat it and enjoy it. I don't eat dairy products now. Mammalian milk contains a substance called opioids. As the name suggests, it is addictive and causes cravings for more. The purpose of this substance is to insure that the infant will stay with its mother for food. In making cheese, the casein is concentrated. It is in casein that the opioids are found. Eating cheese makes you want to eat more cheese. I feel much better, have better endurance, and can keep from gaining weight by eating this way. I eat about every three hours. If I need a snack and don't have time for preparation, I have a food bar called the ProBar, developed by the Food Coach, Art Eggertson, who is actually my food coach. The ProBar is all raw and just good food, made with nuts, seeds, and fruits. It's well balanced and delicious tasting too. You can order it on our website www.secretsofthefountainofyouth.com. My food coach tells me that as we age if we eat fewer calories we improve our health and our longevity proportionately. If we don't eat fewer calories as we get older, eating and exercising the same, we will gain about 10 pounds every decade after the age of 30. I should know; I trained up to four hours a day and still gained weight. It doesn't seem fair. I wonder what the lesson is in that? I am such a chow hound and when I was young I could eat what I wanted and never gain weight, because I always was into sports and exercise. It just doesn't work that way now. Eat fewer calories as you get older; make sure they are dense with nutrition.

If you haven't already started your youthing journey, commit now. Take on one aspect. Engage in it fully. The results will give you the trust and momentum to continue. Each of the chapters in this book is written so you can review it and put it into the plan in a powerful way. Very quickly go back and review the main points of each chapter. Especially review the one on the brain. Transformation takes place in your brain first. You must first choose to think a new possibility about aging. Another way to say aging is "being set in your ways." If you are willing to not — already know or be right about what you have always thought about aging, you can just set up a series of goals you want to accomplish from each of the chapters. You can declare them to your brain daily. Make a visual display also. The more real you make your goals, the greater the chance you will actually fulfill on them. Get a partner on this journey. If you don't have one close to you, at least talk or email your goals, daily actions, and accomplishments to a distant friend, preferably one who is doing the program and is on the journey with you. This empowers you to stay on the road to youthing every single day for the rest of your life.

You can break free of the patterns your brain has been generating regarding the myth of the decline of aging. There are many ways to do that. Here is some food for thought about the subject of aging that may help you break up the mindset commonly accepted as *normal* that would have you "succumb to the inevitable" as you age. Be committed to excellence and extraordinariness. Striving toward extraordinariness is not perfectionism, by the way. Extraordinariness causes new realities. You are willing to pioneer a new way of thinking for the rest of us. A perfectionist will become old fast. Perfectionism is rigid and obsessive within the same system. Someone who has to have life be *exactly* a certain way will not be able to generate the freedom to think outside the current common culture of aging. What you are going to do is get rid of all those set ways of thinking and doing. You are open to the possibility of youthing. You have *chosen* to not participate in the decline of aging. You choose youthing. You are creating a new reality for yourself and others. You become a new model of a new possibility of aging. You are the possibility of youthing for the rest of your life, for the rest of the world, and for the rest of time.

I look for foods that are rich in antioxidants. Herbs and spices have more antioxidants than berries. Cinnamon is at the top of the list. The rule of thumb is the richer the color of the food, the richer the antioxidant content. Shopping for color is a great guide. I grow my own herbs in my green house so I can chop fresh herbs year round. Raw fresh herbs are great in salads. I like to drink a roasted green tea called Hojicha. It can be found in most health food stores. Green tea is a good antioxidant and naturally low in caffeine.

The Best "Carbs" for Your Body

The best carbohydrates for your body are complex carbohydrates or food in its natural state, cooked or uncooked.

Starchy Complex Carbs
All Whole Grains
Dried Beans and Peas
Sweet Potatoes
White Potatoes
Winter Squashes
Corn
Butter Beans
Cracked Whole Grains

Fibrous Complex Carbs
Sprouts
Green Leafy Vegetables
Broccoli
Cauliflower
Cabbage
Eggplant
Salad Vegetables
All Other Color Vegetables

Eating principles:

1. Eat six small meals a day.
2. Eat portions only as big as your fist.
3. Eat something raw at every meal.
4. Use one to two tablespoons per day of unheated flaxseed or olive oil.

The new youthful you has to look for ways to break up those old patterns and those expectations you have hard-set in your life. It is either your left brain or right brain that is having you do things a certain way. You can increase brain flexibility by doing some every day things differently. It is going to be very uncomfortable at the start, but hang in there and things will actually begin to shift in your entire life. Try taking your watch and switching it to the other hand and wear it there until it feels comfortable. This is a way of breaking that pattern. Every little thing you do will help and when you write you should write a little with your other hand, even though it's uncomfortable, just to break the pattern. When you eat, try eating some with your other hand. When you brush your teeth, every other time, brush your teeth with your other hand. If you are left-handed, do some things with your right hand. If you are right-handed, do some things with your left hand. You can shave, you can do your makeup, you can do every single thing you normally do with one hand with the other hand from time to time and break those patterns. It will jar your reality. If you go the same way to work every day, take a different way. If you drive the same car every day, borrow somebody else's car for a day. Notice all the rigid patterns you have established in your life. Do you know losing a job or changing jobs has an impact of upset in your life like having a death in the family? Try doing many daily tasks and routines in entirely new ways. It will support you in thinking in new ways about all your life. These are the things hold you captive and keep you being set in your ways. They also facilitate thinking that is going along with the illusion of the decline of aging. This is just an exercise to do over the next little while to break up the patterns of the brain and get your brain looking at life in a whole different way.

I actually devote a whole day, once a week, to doing these things. Every Wednesday is devoted to doing something different. I call it "Do It Different Day." Here are some more suggestions:

1. Take a trip somewhere even if it's just around your neighborhood. Really look with new eyes. Go down a street you have never been on, take a trip to a town near you that you have never been to, and take a big trip to somewhere you have never been. Go somewhere where you have always wanted to go, yet have just never made it happen.

5. If you eat meat, eat only skinless chicken or turkey breast, and fish — including salmon, often thought to be too fatty, but actually rich in the right kind of essential fatty acids necessary for healthy cell function. Eat only one portion of meat daily. Do include beans with whole grains and tofu with whole grains. If you like, include nonfat dairy products, and low-sugar protein shakes.
6. Eat three to four portions of starchy carbs daily and as many servings of fibrous carbs as you want.
7. Limit sugar and salt intake.
8. Eat only two to three portions of fruit daily. Drink no fruit juices.
9. Drink 6-8 quarts of pure water daily.
10. Drink no carbonated drinks.
11. Drink no alcoholic drinks.
12. Do not smoke.

Eliminate processed refined foods. Refined processed foods are easily converted to sugar and ultimately to body fat. The body reacts to these foods almost in the same way it reacts to simple sugar. They stimulate an insulin surge that sets off a series of fat-producing reactions. The body uses complex carbs more efficiently. The glucose in them is released more slowly and is stored as glycogen for the body to more effectively use as energy. When you eat more complex carbohydrates you will not crave sugar, especially at night, when it will make you fatter. You will have more sustained energy and will begin to look leaner, even though you may be eating more food on this program. You'll notice the difference in just a few weeks. Processed foods also contain many potentially harmful chemical additives and can cause food sensitivities.

2. Try changing the things you eat. Try saying out loud, "I am the power of my word." Declare you love something you hate. Eat foods you normally don't like, tell yourself you like them, keep eating them until you do. When you dry yourself after you take a shower or bath do the left side first if you normally do the right side first. If you usually get out of the bathtub or shower with the left foot, try getting out with the right foot. Use the towel on your back and change the hand from left-handed to right- handed or right-handed to left-handed. Every time you think of it, reverse all those things in your life. Do all your routines in a different way. If you haven't talked to a friend in many years, get on the phone and talk to that friend. If you are estranged or out of communication with your mother, father, your sister, your brother, or anybody on this planet, call that person up and complete the past issues and reestablish those relationships. Find a way. The whole idea over the next little while is to actually alter the common patterns of your brain and become the master of your own life, the master of your own brain. Use your words, use your declarations, use your goals, use your objectives; and present them visually and auditorily to the brain every day. The brain is the most powerful computer on this planet, more powerful than all the computers in the entire world combined. Take charge of it. You must make your goals specific and tell your brain "you are getting younger every day and you are going to do it by changing your diet, by exercising every day, through nutritional support, eating all the good and none of the bad." Finally, take care of your skin. All these things are extremely important. You will be creating a new you and a new reality.

We are going to review each element of your plan for a new you, one by one. You have the power of your brain and the power of feeding empowering new data to your brain with the power of the word, which is in Chapter 2. You understand that you can start telling yourself today, "I am getting younger every day." You can begin the journey of Merlin and live in reverse and get younger every day. You can know exactly what's ahead of you because you have already been there and done that. You have already arrived. Do you understand that concept? When man stood on the ground and pointed at the moon and said, "We will walk on the moon in ten years," he did not already have the technology to

I used to think that because I was a world-class athlete I needed a lot of protein. I now eat 75 percent complex carbohydrates, 15 percent protein and 10 percent fat. Overconsumption of protein causes ammonia to build up in the body, which becomes a tremendous burden for the liver and kidneys. Too much protein in the body also causes the body to be more acidic and a more acidic environment causes us to be more prone to disease. Maintaining proper pH is one of the body's fundamental requirements in maintaining optimum health and performance. Cancer loves an acid environment. Acidity is also associated with calcium depletion and subsequent bone loss. I'm 63 years old and my bone density is 25 percent above normal for a 25-year-old. Diet is the most significant element in maintaining proper pH. The condition of chronic acidosis also results from eating refined sugar and drinking soda, coffee, and alcohol. Smoking also causes acidosis. Is it any wonder that cancer rates are epidemic in our country? A plant-based diet is the best way to maintain proper pH in the body and also yields the lowest risk of heart disease. One nutrition study showed that people who adopt a plant-based diet reduce their saturated fat intake by 26 percent and achieve a significant drop in cholesterol levels in six weeks.

Along with good natural whole food, I drink sufficient pure water. I have pure spring water coming out of the tap. I think it will make a big difference in our health just to live where we live. If you don't have pure, natural spring water coming from the tap, I suggest you get a reverse osmosis machine. You can check our website for information on reverse osmosis machine. Your body also won't let go of fat if you don't drink sufficient pure water. If you want good health and you are on city water, you will also have to get a water purification system for your shower. Check our website for information on shower filtration. Why bother do that. The *speaking* called forth the *inventing*. In the speaking we were already there. On your journey you can use your words and the power of your words to declare different things about your life. If you have never had enough money or enough of anything, declare you are rich. Declare it every day. If you can believe it, you can achieve it. Do the things that will give you the power to overcome feeling like you are getting old. It's not okay to grow old. It's not okay. Above all else, don't grow old gracefully! Don't get old at all. Be young all your life; be vital; be in charge of your life. That is the first part of transforming your brain. Have the audacity to do these exercises and set specific objectives in each of the areas from the chapters of this book.

You must "program" your created objectives into your brain every day. "Whatever the mind of man can conceive and believe, it can achieve." Go back and read *Think and Grow Rich* and start applying the principles of getting rich to youthing. Read *One Minute Millionaire* and apply the principles to youthing. Write your goals. Actually say *by when* you are going to be younger. Say when you are going to have a body with which you are satisfied, by when you are going to be vitally alive and free of any symptoms of illness, and say by when you are going to be living the life of your dreams. All these things are within your reach. Know you can do anything you want and the brain is the tool and the key to making you younger, healthier, and stronger than ever before. Declare you want to look and feel young again and have the vitality of youth. You can have a life you love and do all the things you want to do. Our motto is "to be physically able to live your dreams." Of course you can be mature, of course you can deal with your children or grandchildren without acting like their sister or brother. We are talking about a new youthful appearance, vitality within your heart and within your mind and within your body.

Take all these elements and form them into a real plan of action. The plan that will emerge will amaze you. Write down each of your goals and read them to yourself every day. Read your goals to yourself every morning and again before you sleep. I know a very successful guy who recorded himself speaking his goals and dubbed in music. He played the music softly in the background in his office eight hours every day. He made a collage of pictures, representing his goals and dreams, to purify your drinking water if every time you shower you absorb more chlorine through the pores of your skin than if you'd had 30 glasses of unfiltered tap water? Water makes a difference to human health. Being responsible for having sufficient pure water is fundamental to good health and is not an option. I don't drink sodas, period. A live blood slide preparation was done of my blood right after I'd had a diet coke. Within minutes the red blood cells began exploding and disintegrating before my very eyes. When my teenage son saw this, he also stopped drinking sodas…"cold turkey." Drink pure water.

"To supplement or not to supplement, that is the question." I supplement. I will read something about not needing supplements if you have a good diet, but every time I stop my supplements, I get a cold or flu. When I supplement, I never get sick. I don't have time to get sick and I don't like how it feels. I supplement. I take the vitamin/mineral formulation put out by the Synergy Worldwide Company that makes the ProArgi-9 supplement that stimulates your own pituitary gland to produce the human growth hormone again. I've already discussed the fact that ProArgi-9 made the biggest difference in my energy and endurance of any supplement I have ever used. I take an omega3 formulation, a bone formulation, and a cartilage formulation also marketed by Synergy Worldwide. Their formulations are high quality and leading-edge technology. I like buying everything at the same place when I can. Take supplements when you eat food so they will be utilized better. The formulations I take are all capsules, not tablets, making them much more readily absorbed by the body. I also take Synergy's SynerFood capsules, which is a good "green" formulation. Again you can go to our website www.secretsofthefountainofyouth.com for more information and how to order these products.

that he cut out of magazines and mounted them on a bulletin board right above his desk. There is power in demonstrating your goals in the physical domain. Writing down goals makes them more real and displaying them makes them more real. Speaking them and listening to them daily makes them even more real. Your mind becomes immersed in your goals and perceives them as real. When the mind perceives the goals as real, it generates them into reality. I recommend the very first goal you set be about your body, your health, and youthing. Without your health you have nothing. I suggest your second goal be about daily study to continue to develop your brain. Life is about learning. Never stop learning. Next, I suggest your third goal be in the area of your spiritual wellbeing; the fourth goal might be to take on planning your precious time here on earth. Make goals about the most important people in your life. Include goals about money and finances. Can you see how these goals can cross over and impact other goals? If you are scheduling your time, make time to create and be responsible for your finances. Plan that weekend, second honeymoon with your spouse. It's your life. Plan it, engage in it, and make your life fully functioning. Set up these goals using "Think and Grow Rich" as the basis for all the goals you have in youthing and take them and apply them and read them to yourself every morning and every night before you go to sleep. Formulate this today — don't wait!

You should have specific goals about what you eat and how much you eat. You know now most of the secrets about food and what it can yield to you in the form of youth. Mastery is what you want to achieve. Reread Mary Louise's chapter on mastery and apply it to this all important part of your life. We have talked to you about your skin the things you need to eat and drink in order to make skin and body more youthful. We have given you all the information about fruits and vegetables and all the good things for your body. Make those a part of your goals and declarations and transform your life, if you are sick, take the monosaccharides to get well. We highly recommend these products to you. Use them and watch your health return to you like you never thought possible.

Start from where you are. Improving your health will allow you to get to the point that you can exercise. Exercise is one of the keys to life

The next fundamental requirement for extraordinary health and performance is exercise. One of the reasons I like the martial arts is that it uses isometric toning when I am doing forms and maintaining stances and positions. It also incorporates aerobic conditioning in practicing technique and anaerobic conditioning with kicking and sparring. I also like the fact that it exercises and strengthens the mind and spirit along with the body. Whatever sport or manner of exercise you choose, you have to sweat. I am convinced that sweating is fundamental to health. Drink lots of water and sweat. I average two hours of intense exercise a day and when I have a major competition coming up, I may bump it up to three or four hours. For me, I found that I make greater gains keeping my training to two hours a day. I don't burn out or feel stressed out on a steady regimen of two hours a day.

I begin my training with active stretching. An active stretch is achieved through movement. Passive stretching is done by holding the body in a position at the limit of your flexibility. Passive stretching should only be done at the end of a workout when the muscles are well heated. With my active warm-up stretch, I start at the top, gently rotating my head, then the shoulders, move on to rotating the torso, the hips, and then the legs. The movement should always be flowing and smooth, never with a bounce. Bouncing can produce tiny micro-tears in the muscles, causing scarring and ultimately less flexibility. There are two reasons why people strain or tear muscles. One is improper warm-up. Active stretching for 15 minutes is a proper warm-up. I use active stretches that also include muscle strengthening and balance development, getting "three for one" in my warm-ups. The other reason to strain or tear a muscle is putting too much load on a muscle before it is strong enough to bear that load. You must strengthen slowly over

— not just to youthing but to living a life you love. Exercise allows you to experience natural highs through endorphins. Exercise enables you to get your body under control and do the things the body was meant to do. Cut down on the amount of food you intake and make changes to the things you eat. Get sugar out of your diet; get rid of anything that will cause you to be overweight. Never, never, never give up. Go back to the charts and set a goal to be at or below your ideal weight. Live the rest of your life at that weight. You can do this. Your brain will help you and your friends will help you. We have a system set up that will allow you to have a life you love and get back to where you truly want to be. You will have a body that can do those dreams and goals for the rest of your life.

Correct nutrition is vital. We have found that vitamins, minerals, and natural supplements absolutely make a difference. We are defying the decline of aging and have blazed the trail for you. We believe the most powerful supplement on the face to the planet today is ProArgi-9. It literally says to the body "I am getting younger every day." It reverses the aging process, it has your body naturally produce things the way it did when you were in your youth. It allows you to exercise and to get your weight back under control. It frees your brain up and allows you to do all the things you did in your youth. It gives you vitality. ProArgi-9 is one of the most important secrets of the fountain of youth.

Consider some of the other secrets we presented. We told you about Mistica Juice, goji juice, noni juice, pomegranate juice, and adding your own fresh fruit and vegetable juices to your diet everyday. Read Jay Cordage's book on juicing, or read any other books you can get from the library and start juicing every day. Start bringing the enzymes and the living things back into your body. There are the fundamentals of nutrition for health. Transform the brain by setting goals and using the power of your word to just do what works to have a body you love.

A goal is a goal if it occurs in time. In setting your goals, establish a six month timeline. No matter what shape you are in, no matter where your life is right now, within six months you can have life the way you want it, simply by following the plan you are organizing right here. Schedule your time accordingly; remember, within six months you can have all these things happen. Create a timeline. Then work your timeline time. Ballerinas and martial artists have incredibly strong hip flexor muscles from doing thousands of leg lifts over years of time. I can throw 500 kicks a day but a novice would start with 50. I live in an area where I can't run much in the winter. I prefer snowshoeing to running on an indoor track where I have to travel 30 minutes to get to it. Consequently, in the spring when I begin running again, I have to run slowly the first few weeks, only striding out lightly to improve endurance and strength before I think about doing sprints to improve speed. As a martial artist, I love to jump rope for endurance and quickness. It is sport specific in that I jump for three-minute intervals, which is the same interval of a round of sparring. Again, if I haven't been jumping for a while, I'll start out with 250 jumps and work up to 500 jumps. I begin with two rounds and work up to four or five. The other great part about jumping rope is that a good jump rope costs only about $20 and you can do it anywhere.

After my active warm-up, I do about 200 leg lifts, lifting to the front, the side, and the back. With a front leg lift I can bounce my thigh off of my chest now, but when I started 20 years ago, I could only lift my leg waist high. Next, I do 100 slow-motion kicks, front kick, back kick, side kick and round kick. I often use weights with my slow-motion kicks, because the purpose of slow-motion kicks is not only to develop proper technique through repetition, but also to develop muscle and ligament strength. Of course you wouldn't use ankle weights until you could comfortably do ten repetitions with each leg and each kick. Never use ankle weights to do kicks at full speed. You might injure a muscle, but you will also make your kicks slower. Trust me on this one. They've done research. Use weights only while doing slow-motion kicks to improve strength. When strength improves, so does speed. Doing high repetitions every day. Take charge of your schedule and take charge of you. One of my favorite spiritual sayings is, "Cease to be idle, cease to be unclean, cease to find fault one with another." Plan to get 8 hours of sleep every night. "Arise early that your body and your minds may be invigorated, retire early that ye be not weary." Nothing is more important than scheduling your life for success. If you are a night time person, change that and say, "I am an early morning person." Be an early riser. Early morning is rich with joy and energy for the whole day.

Follow these guidelines and change your routines. Make sure things are happening daily. One of the essential elements of success is to enroll other people in your project of a new you and a new life. You enroll other people by sharing your project with them. Perhaps you'll enroll them in taking it on with you and you'll keep each other going. Let's briefly review each chapter. Remember if you want to master each chapter, go back and review the whole chapter and apply the principles until they are a part of your being and you experience a transformation in that area.

1. Chapter 1 — The Brain

Wake up your brain to transform your life by doing things differently.

2. Chapter 2 — The Word

Own the power of your word. Remember, you are your word. Set goals. Design a plan of effective action. Work your plan. If you want to rehabilitate the power of your word, do "Landmark Education's Curriculum for Living." It works. Register for their first course, "The Forum," today.

3. Chapter 3 — Declarations

Make powerful declarations. "I am getting younger every day." Organize your words into powerful statements that will wield the power to transform your life into a life you love.

4. Chapter 4 — Health and Wellness

Be well. If you are ill or have any condition that stops you from youthing, reread this chapter and start taking the nutritional supplements and actions that will support your body in regaining health and wellbeing.

5. Chapter 5 — Weight Loss, Fat and Muscle Balance

of full-speed kicks on a bag also improves speed.

After my slow-motion kicks I do about 300 full-speed kicks. I'm sweating profusely at this point. Then I practice my competition form for about 30 minutes, followed by some passive stretching. Later that evening when I teach, I do footwork, sit-ups, push-ups, and abdominal exercises with my students. Two days a week I also do weightlifting. I actually hold the world record for the chest press in women of my weight and 60 to 65 years of age. I don't love to run or do weights. Some people do love it. I love to do Taekwondo. It enlivens me. Everything else I do so I can be my best in Taekwondo. After 20 years of this…I'm strong and fit. The human body is not the limit. It will continue to strengthen as long as you do proper strengthening and conditioning exercise. The flexibility I enjoy, the strength, balance, and speed I have are not because I am extraordinary. I just do the work. If everybody did the work, it wouldn't be extraordinary. It would be ordinary to be well, strong, fit, and able to do our dreams all the days of our lives, not just when we are young.

Russian Adaptogens: Nature's Phenomenal Secret for Extraordinary Health and Performance

Stress and stress-related disorders contribute to perhaps 75 percent of all illness in modern man. When under stress, the body reacts by producing adrenalin, which stimulates the whole body to a hyper-alert state. Metabolism, blood pressure, and circulation are disrupted, immunity and resistance are drastically reduced, and performance suffers. In short, stress destroys well-being, damaging health and performance. Low energy is a natural fallout of high stress. Western medicine has recognized the significant role of stress in the etiology of illness and has attempted to cope with

You can create a new you with a balance in life that is achieved only through applying the principles in this chapter.

6. Chapter 6 — Exercise and Finding a Trainer

Start your exercise program and you start your life. Find a trainer and you are on your way.

7. Chapter 7 — Food and Nutrition

There are foods and nutritional supplements that can take you to the next level of life. You need to know what they are and experience the miracles that come through proper nutrition and breakthrough supplementation.

8. Chapter 8 — The Skin

Study this chapter carefully and just try the skin care protocol we have discovered that has people guessing our age to be twenty years less. Bring a new youthful look and feel to your skin. Treat your skin internally and externally. Give it the care it needs by setting up a daily routine of hygiene and health both morning and night.

9. Chapter 9 — The Plan

Set your goals. Make a timeline. Make a plan of action. Work your plan. Try changing your daily routine. You can have a life and a body you love. You can be youthing instead of aging. Enroll other people in youthing. Enroll them and they become your partners. Tell them your goals. Tell them what you are up to. Tell them what you feel and what you believe. Ask them to support you in your new life. Have them help you to change and to have a life you love. A new life is what you want. You want a life full of vitality and energy. You want it for yourself and you want it for others. You can enroll your friends at work, anybody you talk to, and even a total stranger. They can help you live up to the commitments you have made to the new, younger, vitally alive you.

10. Rest and Relaxation. The ability to rest is extremely important. We believe in getting plenty of sleep, at least eight hours a day. Make sure you love yourself enough to do this and to get adequate rest every day. Rest and relaxation are a big part of youthing. You need to take the time to take care of yourself, to meditate, to go places where you have never been before, to take the time to relax, and to enjoy every moment of life. Rest and relaxation are some of the keys that will help you achieve a new life. Take the time to relax. Hawaiians use the phrase,

stress with several approaches, including pharmaceutical drugs and exercise and relaxation techniques such as meditation. Pharmaceuticals can help with the management of the symptoms of stress without really protecting the body from the damaging effects of stress, and often have negative or even dangerous side effects. Exercise can be effective in reducing the damaging effect of stress on the body, but for the competitive athlete, the intensity of four to six hours of daily training can be another source of serious stress for the body. Meditation has been proven to also reduce the damaging effects of stress on the body and yet most people don't apply the daily discipline needed in order to benefit from meditation.

In Russia, researchers also struggled to find solutions to the physically damaging effects of stress and after years of scientific investigation, developed a unique approach to stress reduction and the prevention of stress-related symptoms and illnesses. A leading Russian pharmacology professor, Lazarev, believed that the answer to improving resistance to the ever increasing stresses of modern life lay not in treatment with pharmaceuticals, but in protection and prevention with neutraceuticals or plant phytonutrients. Over 200 Russian scientists engaged in more than 20 years of research to prove the efficacy of phytonutrients in human health and performance. Among these scientists was a renowned Russian research pharmacologist, Dr. Israel Brekhman, who coined the term "adaptogens" for this classification of elite herbs. An adaptogen is a type of plant with certain characteristics. First, it is absolutely safe and nontoxic, producing no negative side effects. Second, it increases the body's nonspecific resistance to disease, strengthening the immune system. Adaptogens have demonstrated a wide range of therapeutic effects in influenza, hypertension, diabetes, atherosclerosis, obesity, "talk story." What it means is taking the time to just sit down and be with people and talk about life or anything else you want. You can find out what friends are up to and learn to relax. Now, you can choose to "talk story" and change Your story, that you are too busy. It is not a good thing that everybody in this world has somehow unfortunately misinterpreted what being busy is all about. Some people think it's a good thing that you are busy every minute of every day, you don't have time (for anything). You are too busy running around, to be with people. That is absolutely not true. One of the most important things you have to schedule is to sit back and relax and to understand life doesn't have to be a blur. Life just doesn't have to be that busy. You can take the time to stop and be involved with other people, with their lives and have them involved in yours. To embark on the journey of youthing, you must sit down every day and "talk story" with someone.

11. Youthing is an every day thing. Say to yourself, "My life is worth living." I choose to make the changes offered to me here. I will become younger every day through the actions I choose to take. I generate a wonderful life that I love. Mastery happens a day at a time, a choice at a time, an action at a time. We invite you to be our partners in creating a new culture of aging, one of living life vitally alive all the days of your life. Youthing is a journey that starts with the first step and is enlivening all along the way. You will never arrive anywhere. You will be on an endless adventure from the first moment to the last. Life is great. The more alive you are the greater your experience of life.

12. The Website. We have dedicated our website to helping you find and implement all the secrets of the fountain of youth. Anything new we find, anything we think will help you in any way on your youthing journey will find its way onto our website. It is the greatest source of youthing there is. Everything current about youthing, everything about nutritionals, everything about any type of new products related to youthing will be on the website. Your key to becoming younger every day will be found at www.secretsofthefountainofyouth.com.

We will be releasing new books in the future. Mary Louise is writing a book called *The Ninja Grandma*. She says her purpose in writing that book is to inspire you to take on doing your dreams no matter what your age. *Secrets of the Fountain of Youth* is being translated into and cancer. Adaptogens work generally on all body systems to normalize and balance body function. They are "anti-stressors," increasing the body's tolerance of stress and endurance in the face of stress, both physical and mental. In the '50s and '60s, massive clinical trials were conducted in the USSR. The findings showed that plants with adaptogenic properties not only improved all body system functions and strengthened the immune system, but also dramatically increased energy, stamina, strength, speed, but most importantly, they enhanced resistance and adaptability to all kinds of stressors. Adaptogens have been proven to decrease the harmful and damaging effects of stress on the body.

The Soviet government realized that this new class of phytonutrients, adaptogens, could give the Soviets an advantage in sports, space and military programs, and consequently accelerated and supported this scientific project for decades. In the '70s, information about this remarkable new class of herbs with adaptogenic properties began to spread around the world. Soon research scientists in Germany, Sweden, Japan, and the United States began to confirm the effectiveness of adaptogens in their own research. Some of the latest scientific evidence shows that adaptogens work at the cellular level to increase the body's ability to manufacture and use cell fuel more efficiently, even in periods of intense stress. Adaptogens also act as antioxidants, protecting cellular membranes from the adverse effects of free radical buildup. Adaptogens also have an anabolic effect. They aid in rebuilding damaged muscle tissue following strenuous exercise. This is accomplished by facilitating the synthesis of protein in the body. When supplementing with adaptogens, the time needed for the body to recover from high-intensity training or extended periods of stress is greatly reduced. This increase in anabolic

Japanese, Korean, and Spanish. We will release sequels to *Secrets to the Fountain of Youth* with new secrets from all over the world. Secrets of youth that will come from exotic places like Europe, Africa, Japan, Korea, Latin America, and China. All the information about these new releases and products and how to get them at a substantial discount will be on our website. Everything you need to continue on your youthing journey will be at your fingertips at www.secretsofthefountainofyouth.com. Don't forget to sign up for our newsletter, it will have exciting new information in every issue. Our website can become your escape, your personal oasis. We highly recommend you go there and go there often.

Aloha and "bon voyage." May your journey be a enlivening one.

activity is achieved without any negative side effects.

The term "adaptogens" has become almost common today and there are many adaptogenic products on the market, but not all adaptogens are created equal. A recent study that evaluated so-called Chinese and Korean ginseng products and many other adaptogenic formulations, found most of them to have limited or no value. This is due to poor plant quality, incorrect species, incorrect harvesting times, over-processing of raw materials, and ineffective technology in extracting active substances from the raw herbs. Most producers of adaptogenic formulations refer to Russian research to market their products, but have no knowledge of their patented six day-cold extraction process to produce the superior product and effect. Nor do they have access to the primordial region of Russia where these rare plants have been growing for millennia. Scientists have found in the chemical composition of an adaptogen, Eleutherococcus senticosus, from this region in Russia large amounts of eleutheroside-B, which plays the most significant role in stress protection and stimulating anabolic activity. In China, eleutherococcus is known as ciwujia. It is from the same botanical family as the Russian plant, but ciwujia contains little or no eleutheroside-B.

Dr. Ben Tabachnik, PhD, a former senior Russian scientist and head of the research group for the Soviet Olympic team, supervised leading scientists in the areas of psychology, physiology, biochemistry, pharmacology, and nutrition. Their work was classified as a "state secret" just like the military and space programs. Their goal was to make Soviets athletes the best in the world. Using the research of the hundreds of Soviet scientists working with many different adaptogenic herbs, Dr. Tabachnik's adaptogenic formulas were developed and tested with Soviet Olympic athletes. The Soviet record of 55 gold medals

CHAPTER 10

MY PERSONAL SECRETS

For Your Eyes Only

Now I will tell you about my own daily youthing regimen. At age 57 I look, feel, perform and run like a 35 year old. All I have learned through studying and my training as a coach and through all the different fitness programs and workshops in which I have participated comes down to a few core truths. Fundamentally, it is very simple. I want to focus on the word, *core*. It is a word that is currently a *buzz* word in the fitness industry. The popular use of the word originated in the disciplines of Pilates and Yoga. *Core* refers to the central area of your abdomen and is purported to be the center of all your physical activity and performance. I believe the *core* exists for all three areas of your life — physical, mental and spiritual. All three of these domains have a core. The core of your physical body is your abdomen area. In strengthening your *core* muscles, you will improve all your physical performance. The core of your mental capacity is your brain, which includes your thoughts. You must work equally hard at keeping your mental core in shape. Finally, you have a spiritual core as well which is represented in your heart and can be strengthened through love, service and communion with God.

in one Olympic Games has never been equaled. Rumors flew! People thought that the Soviet athletes were surely on steroids, but they could never prove it. They weren't! They were on adaptogens, which when tested, have never revealed any banned substances for competitive athletes.

When Dr. Tabachnik emigrated to the United States he began working with an American company as the vice president of research and development. His formulations were successfully used by top-level athletes from Russia, the United States, Australia, Germany, England, and Canada. For the last ten years Dr. Tabachnik has been working to develop the next generation of adaptogens, reflecting the increasing demands of modern society. Like drugs that saved people of the world from bacterial epidemics that in the past took millions of lives, adaptogens are now needed to help man survive the intense stresses of today. There are no other products in the world that have the quality and potency of the Russian adaptogens and now not even in Russia, that have the effectiveness of Dr. Tabachnik's formulations.

While in Russia, Dr. Tabachnik published more than 150 articles and books on enhancing sport performance. Many of his writings have been widely reprinted internationally. Shortly after his immigration, Dr. Tabachnik published his first book in English, entitled *Soviet Training and Recovery Methods*. He immediately established himself in the United States as an expert in sports training and performance by consulting with professional teams such as the Denver Broncos, Detroit Pistons, and Chicago Bulls. He has also been active in training many elite professional and collegiate athletes. Dr. Tabachnik has been a guest on numerous talk shows and has been a featured speaker at many conferences and clinics. He has been recently recognized in the world of herbal medicine after his presentation at international

	Area of the Body	You Work on Your	What to do
Physical Core	Abdomen	Muscles	Exercise
Mental Core	Brain	Thoughts	Study & Read Goals
Spiritual Core	Heart	Feelings	Love & Serve Others

If you are going to be on a journey of youthing for the rest of your life, it will include strengthening the core of all three areas. Mastery comes to those who persevere. There is really no arrival. Mastery is endless. The more you know about something, the more you know there is more to know. Life, mastery, and youthing are all never-ending journeys.

Each day I follow certain regimens in my life. I arise at a regular time, I follow a routine that is disciplined in many ways but, I also try to be flexible and am always adding new routines. It prevents the old age malady of being *set in my ways*. I will share my regimen with you and hope that it will make a difference for you. You may not agree with all my beliefs or philosophies, and feel free to use whatever works for you. You may start out borrowing something from me and building your own style and philosophy. Once you are on the path, youthing will continue to develop through you. Youthing is alive and growing.

Each day I arise at an early hour. I practice the old adage "the early bird catches the worm." I believe arising early invigorates the mind and the body. I believe retiring early makes me less weary.

My Day

I arise between 4:30 a.m. and 5:30 a.m.

I usually wash my face, brush my teeth, and shave to start the day.

Then I immediately go to my office and start my day off with a prayer. I believe in God. I was raised a Christian and have strong faith in Christ.

I then read scriptures for at least one half hour.

Next, I study motivational books for another half hour.

Then I read my abundance statement, "I am abundant in every good way. Infinite money is mine to generate, earn, save, invest, exponentially multiply, and share. My abundance is making every one better off. I embrace abundance and abundance embraces me."

symposia on the use of adaptogenic herbs. These talks included a conference on the "Crisis of the Immune System" in Israel and the "International Conference on Stress" in Switzerland.

Dr. Tabachnik is now also working with Dr. Stephen Fulder, graduated from Oxford University and a world-recognized expert in herbal medicine, molecular biology, and anti-aging. Dr. Fulder has written 15 books and is coauthoring a book with Dr. Tabachnik entitled, *Fighting Stress with Adaptogens*. Dr. Tabachnik's newest formulation is an adaptogenic product designed as a universal supplement that can be considered to be a strong foundation for any nutritional program for stress management, sport performance, longevity, weight control, disease prevention, and the promotion of extraordinary health and well-being.

I was one of the first five people in the United States to take Dr. Tabachnik's formulation of Russian adaptogens. My husband, Ron, and our son, Adam, then five years old, soon joined me in using Russian adaptogens as a daily supplement. While on our daily regimen of adaptogen supplementation, my husband and I went ten years without having a fever or any illness at all. Maybe that could be explained by our already good diet, supplementation program, and consistent exercise, but I never heard of a five-year-old reaching 15 without having a fever or a cold or flu. Adam had many ear infections as a toddler and normally had as many colds and flu as other kids, as did we, until we started on adaptogens. Being an older athlete competing with people less than half my age, I now had a "Soviet secret" of my own. Adaptogens dramatically increased my endurance, strength, speed, and recovery. It was literally a godsend. I take many leading-edge supplements, but adaptogens were pivotal in my supplementation program.

After that I write each of my six goals in my journal. I read those same goals five or six times every day and once more just before I go to bed. I have one goal for each of six areas: Body, Brain, Being, Time, People, Money.

I have a business meeting every morning with my business partner at 6:30.

I then go out for a run every day except Sundays. I am a marathon runner. I am running the Boston Marathon again this year in April. I have qualified for the Boston Marathon almost every year for the past 20 plus years. Typically, I run an average of 10 plus miles a day and 15 to 20 miles every Saturday. I still go to church every Sunday and since I honor the Sabbath, I don't run on Sundays. Before I run I take one or two supplements depending on how much I am running that day and how I am feeling. Whenever I am doing 10 miles or more, one half hour before I run I take one scoop of ProArgi-9 mixed in with one scoop of Accelerade. Accelerade is a relatively new discovery for me. I read a book about the most famous bicycler in the world. He uses a 4-to-1 ratio of carbohydrates to protein called Accelerade. This unbelievable combination yields a whopping 29 percent increase in endurance and the darn stuff works. When I combine Accelerade and ProArgi-9 before I do long runs I am truly good to go. Find out more about my unbelievable pre-exercise cocktail on the website. It is absolutely the best pre-exercise combination in existence.

After my run, I usually shower, shave, brush my teeth, and get ready for the day.

My morning routine includes using the facial products I have already mentioned. I use Renewing Facial Scrub twice a week as recommended. It is scientifically formulated and is the best exfoliant on the market. It leaves your face feeling perfect. It is "mildly" potent. I use Delicate Face Cleanser every other day in the morning and every night. Delicate Face Cleanser is the best-feeling thing you have used to cleanse your face. It is so mild that sometimes I use it several times a day. I can't even describe it; every single person who has tried this amazing product feels the difference it makes. It leaves your face feeling young and smooth like a baby's bottom. You have to try this. I then follow up with Revitalizing Toner for pH balance. This leaves your face feeling fresh and I really mean fresh. Then I next use DNA Repair Crème. This cutting-edge product lives up to its name. I apply this

Adaptogens gave me a body that could do my dream of being a world-class athlete in my 40s, 50s, and now in my 60s. I traveled around the world to train and compete in Taekwondo. Unlike everyone else traveling through time zone differences, such as nights and days upside down, I had no jet lag. Sometimes, after training all year long, an athlete is knocked out of a national or world competition with the flu. I was never sick or stressed for my competitions. Adaptogens also significantly improve focus and concentration, which is an essential part of successful performance. Strength of focus is as important as practiced skill in performance. Adaptogens have yet another important application based on the improvement of focus and that is with children and adults who struggle with attention deficit disorder.

More importantly, my husband, Ron, was diagnosed with stomach cancer the week before he went on adaptogens. He had a lymphoma primary to the stomach, a highly aggressive tumor. He chose not to follow his doctor's recommendation of surgery, chemotherapy, and radiation. In six months he had no sign of the disease. He did also go on a raw, organically grown vegetarian diet and quit his high-stress job, so it's hard to say just what was responsible for his spontaneous remission. It was probably all of it, with a bit of miracle thrown in for good measure, but the Russian research did show that adaptogens had both a preventative and restorative effect in people with all kinds of cancers. Whatever it was, we'll take it! It's been 13 years and Ron still has no sign of the disease and at 74 years old is still a world-class business consultant and seminar leader, as well as breaking records in running and weight lifting. I just hope none of my competitors in Taekwondo read this book. I am more than willing to share my secret of adaptogens with everyone else.

to my whole face and neck. Whenever I have any kind of problem, like too much sun or wind or any area of my face that seems taxed, DNA takes it all away. I have had very severe sunburn disappear overnight using DNA Repair. I use it every day, and several times a day if I have extra need. I have even found that it causes age or liver spots to fade and skin discoloration to all but disappear. My face loves DNA Repair Crème. I follow that with Advanced Face Cream. This is one of those special products that has the skin produce its own collagen and gives your skin that youthful, soft glow and firmness that is hard to describe. I use this all over my face and neck as well. The final two products work in tandem to help dissipate fine lines and wrinkles. I use Anti-aging Serum wherever I have fine lines. This product visibly works. You can actually see the results sometimes within days. The final product I use in the morning is TriAction Age Defense Complex. You won't believe the science behind this product. The most recently released of all the products, TriAction brings you the latest research on peptides that help to regenerate your skin and counter the visible effects of aging. Each peptide compound performs a different task in aiding you to look your best. TriAction actually helps to noticeably decrease the depth and size of wrinkles and promotes increased skin thickness and optimal epidermal balance and tone. That usually ends my morning routine. This whole process sounds lengthy to say, but only takes me a few minutes. I use Advanced Face Cream and Anti-aging Serum usually in the evening before I retire.

I usually eat a light breakfast before I get officially started on my day. I typically eat a medium sized Fuji apple followed by a banana and usually a half of a fresh papaya. I drink some filtered water and follow up with my supplements. I used to juice fresh carrot-apple juice every morning. Now I have that once or twice a week.

In the morning I usually take my VitaPak, which is the most powerful vitamin pack on the market. I have done my research on this one. I have personally tried just about every vitamin package on the market. I follow that up with coral calcium and a combination of monosaccharides.

I then go to work and have energy pretty much the whole day. I practice reading my goals during the day at least five times. I carry them around with me on a note card in my pocket. On the back of the note card I take notes of anything that pops into my brain during the day. This practice has really helped me put some great ideas to use.

The Importance of Glutathione

Glutathione is a small tripeptide molecule and contains three amino acids, glutamic acid, cystein, and glycine. Glutathione is a compound present in almost every living cell of the body. It must be generated within the cell from its precursors before it can work effectively in the body. Our body naturally produces glutathione by processing the food we eat and manufacturing the three necessary amino acids needed to create glutathione. The liver is the main production and storage site of glutathione.

Cellular levels of glutathione diminish as we age. Diseases such as diabetes, cancer, heart disease, HIV, AIDS, arthritis, immune deficiency diseases, hepatitis, Alzheimer's disease, Parkinson's disease, sickle cell anemia, cystic fibrosis, and male infertility are all associated with diminished levels of cellular glutathione. The lower the glutathione level, the more prone you are to disease. Glutathione acts as a detoxifying agent by combining with toxic substances and ridding the body of them through bile and urine. Glutathione functions as a major frontline defense against the build-up of oxidants in the lungs. It is often referred to as the master anti-oxidant. Higher glutathione levels have also been found in older adults with exceptionally good health. Our glutathione level actually indicates our state of health and can predict longevity. The function of glutathione in our body is so important that in a few years, glutathione levels will be a part of standard physical tests, like blood sugar and cholesterol levels are now.

Increased levels of glutathione have been shown to slow down the aging process, detoxify and improve liver function, strengthen the immune system, reverse the symptoms and

I eat a healthy lunch. Most days I eat a salad. I usually spruce it up with either tofu or chicken. I try to maintain my 4-to-1 carbs-to-protein ratio throughout the day. After I eat, I follow the meal up with a Mistica, Noni Supreme, Synemax cocktail. This powerful combination keeps me going all through the day into the night. I have pure energy and seldom, if ever, get tired.

When I am officially training for a marathon, I run twice a day once in the morning and once in the afternoon or evening. I coach both Track and Cross Country at the local high school. One of the spiritual or community service things I do to try and return to the community the blessings I have received from others.

My dinners are usually fairly light. I try to eat a well balanced dinner with a mixture of carbs and protein. Nothing special here except I always try to include fresh veggies and a little bit of meat to get my 4-to-1 carbs-to-protein balance once again.

I try not to eat after 8 pm and just drink some water after that hour. Approximately, one half hour before I go to bed on an empty stomach, meaning I haven't eaten anything for the previous two hours, I drink a small glass of water with two scoops of ProArgi-9 stirred in. The ProArgi-9 crosses the blood-brain barrier during alpha sleep at night and my body is on it way to youthing. This is the magic fountain of youth. Do this for about six months or so and all your hormone production, especially the HGH, will be naturally produced by your body in exacting amounts and have your body feeling like it did when you were sixteen years old. This stuff really works. Every one who tries it gets the same results. Start this regimen today!

Before I retire I wash my face with Delicate Face Cleanser. I love this stuff. Your face will never feel the same. It feels so youthful! I then apply Revitalizing Toner, DNA Repair Crème and TriAction, Advanced Face Crème, and Anti-aging Serum. Information on all of these products is available on the website.

Just before I go to sleep, I read my "abundance statement" and my six goals to myself. I sleep like a power nap all night. When I arise in the morning, as long as I don't eat any sugar, I am never groggy. I hit the deck literally running and I feel absolutely fantastic every single morning. I am wide awake in an instant and ready to go. Sometimes it even amazes me. Remember that every Wednesday is "Do Something Different Day" to get conditions associated with arthritis, and reduce the chances of developing heart disease and cancer.

Research has shown that individuals who have low levels of glutathione are susceptible to chronic illness. Research also shows that glutathione levels decline by 8% to 12% per decade, beginning at the age of 20. Glutathione levels are further reduced by continual stress on the immune system such as illness, infection, and environmental toxins. This is a vicious cycle. While you need glutathione for a strong immune system, a weakened immune system produces less glutathione.

Unfortunately supplements containing glutathione alone will not increase the body's glutathione levels. It is pointless to purchase supplements that merely contain glutathione, because the digestive system breaks down the ingested glutathione and it will not be absorbed into your system. There is a very new, patented product called MaxGXL, a glutathione accelerator. MaxGXL is distributed by the company Max International in Salt Lake City, Utah. MaxGXL's patented formula has been proven in clinical trials to dramatically increase intercellular glutathione. MaxGXL was created by world-renowned immunologist Dr. Robert H. Keller, M.D., MS, and FACP. I recommend that you "Google" Dr. Keller and you will see he is a giant in the medical field. Some night when you have insomnia, get on the Internet and "Google" "glutathione and anti-aging," or "glutathione and cancer or aids or arthritis", etc. Even though there have been over 60,000 studies related to the benefits of glutathione, the general public is still very unaware of its importance in health and anti-aging. MaxGXL increases cellular glutathione by providing the proper nutrients needed to promote the body's own ability to manufacture and absorb glutathione. It is formulated to get through the digestive system without getting broken down.

you out of your "set-in-your-ways" getting old syndrome. You want to feel young again? Follow the secrets in this book.

One final thing, make yourself accountable to someone. The reason we have the website and coaching made available to you through the website is so that you will have to "return and report." One of the most important things you can do for overcoming sickness and disease as well as losing weight and obtaining your goals is to have someone to account to. If your mind knows you will have to report back to someone, it will believe it and achieve it. Get a coach or a trainer or both!

Mary Louise and I really do have lives we love. We follow the regimen we recommend. If you follow these recommendations I know you will have a body and a life you love too. You truly can have a life you love. This works. Youthing is all up to you. It is a journey. It truly begins with the very first step. Try just saying to yourself five times a day, "I am getting younger every day" and watch your life start transforming right before your eyes. Begin with just that. Then study and practice what you learn in each of the chapters in this book and you will have the life of which you have always dreamed. Your youthing journey will never end if you just say so. Be unstoppable. Recruit your own brain first and then recruit the rest of the world and you will never look back. Visit our website as often as you can and email us your successes. We truly would love to hear from you. Good luck!

Aloha and God bless you; every one.

I can honestly say that MaxGXL is one of the most important supplements in my youthing regimen. Just reading the science will convince you of that, but I could feel the impact in my performance in just two weeks and the knee pain I had been enduring for the last two years was gone in one month. I personally know someone who was crippled by rheumatoid arthritis at the age of 42. After three months on MaxGXL, not only does he walk normally, but his RA values on his blood test went from over 100 to 6. I saw the report. I also have some arthritic and degenerative changes in my vertebrae and the timeliness of my learning about glutathione and MAXGXL will probably be fundamental in my being able to continue to compete in Taekwondo and accomplish my dream of competing in the Olympics. I was sure not looking for one more supplement to add to my regimen, but there is no way I could be committed to reversing the common effects of aging and not take this supplement. Thanks to my Taekwondo student and friend, Jenny Stein, I was introduced to and educated myself about the importance of glutathione and this breakthrough product, MaxGXL.

CPSIA information can be obtained at www.ICGtesting.com
Printed in the USA
LVOW071018170212

269162LV00006B/72/A

9 781933 973043